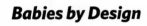

Babies by Design

Ronald M. Green

Babies by Design

The Ethics of Genetic Choice

YALE UNIVERSITY PRESS
NEW HAVEN AND LONDON

A Caravan book. For more information, visit www.caravanbooks.org.

The quotation beginning chapter 5 is from page 3 of *Brave New World* by Aldous Huxley, 1st Perennial Classics ed. (New York: HarperCollins, Harper Perennial, 1998). Copyright 1932, 1960, 1964 by Aldous Huxley. Reprinted by permission of HarperCollins Publishers and by permission of Georges Borchardt, Inc., on behalf of the Laura Huxley Trust.

Designed by Gregg Chase.

Set in Whitney type by The Composing Room of Michigan, Inc.

Printed in the United States of America by Vail-Ballou Press, Binghamton, New York.

Library of Congress Cataloging-in-Publication Data

Green, Ronald Michael.
    Babies by design : the ethics of genetic choice / Ronald M. Green.
      p. ; cm.
    Includes bibliographical references and index.
    ISBN 978-0-300-12546-7 (alk. paper)
    1. Medical genetics— Moral and ethical aspects.　2. Genetic engineering—Moral and ethical aspects.　3. Human reproductive technology—Moral and ethical aspects.　4. Prenatal diagnosis—Moral and ethical aspects.　5. Genetic disorders in children—Prevention—Moral and ethical aspects.　6. Genetic counseling—Moral and ethical aspects.　I. Title.
    [DNLM: 1. Genetic Engineering—ethics—Personal Narratives.　2. Genetics, Medical—ethics—Personal Narratives.　3. Religion and Medicine—Personal Narratives.
WB 60 G797b 2007]
    RB155.G74 2007
    174.2—dc22

                                              2007019927

A catalogue record for this book is available from the British Library.

The paper in this book meets the guidelines for permanence and durability of the Committee on Production Guidelines for Book Longevity of the Council on Library Resources.

10 9 8 7 6 5 4 3 2 1

To Julian and Aniezka

Contents

Introduction

The low beige building with green-trimmed windows in an industrial park west of Sacramento does not look like the center of a revolution. Inside, white-suited technicians working in clean rooms use methods borrowed from nearby Silicon Valley to inscribe millions of microscopic features on blue glass wafers the size of a thumbnail. But the features being etched on the chips manufactured by the Affymetrix Corporation are not electronic circuits. They are stretches of DNA. Using Affymetrix's chips, genetic researchers around the world are now able to perform rapid scans of genes being expressed in the cells of living organisms and almost instantly determine the differences between sequences of DNA. One Affymetrix chip can identify many of the hundreds of thousands of DNA letters (As, Cs, Ts, and Gs) on the human genome that vary from one person to another. Using these chips and the more advanced, information-packed models now being planned, researchers will soon be able to identify the genetic features that distinguish you from me.

The revolution that Affymetrix is part of involves an unprecedented growth in our ability to understand—and eventually change—the human genome. From factories and research laboratories to medical clinics, we are entering the era of directed human evolution. Human beings have always been subject to evolutionary pressures. But for most of our history, we have been the passive subjects of change. In this new era we will take the direction of our evolution into our own hands.

To a small extent, this is already happening. It is commonplace for a pregnant woman to undergo ultrasound diagnosis and other tests to identify—and avoid—genetic or developmental anomalies in the child she is carrying. Thousands of couples that carry a genetic disease and want to avoid transmitting it to their children now use preimplantation genetic diagnosis (PGD), the genetic screening and selection of early embryos produced by in vitro fertilization. Some clinics are already offering PGD to enable couples to choose the sex of their child, the first instance of using genetic selection for a nondisease-related condition. With these and other technologies, a new field of medicine has emerged that unites reproductive with genetic technology: reprogenetics.[1]

Driving these developments is an explosion of information. In April 2003 the Human Genome Project completed the identification of the sequence of three billion paired genetic letters, or "nucleotides," that form the instruction program for every feature of our biology. This sequence was published on the Internet along with the identification of an estimated twenty-five thousand genes, those portions of the sequence coding for the proteins that form our cells and tissues.[2] Accomplishing that task took thirteen years and cost about one dollar for each pair of DNA letters. Thanks to technological advances spurred by the Human Genome Project, the cost of DNA sequencing is now in free fall. Genome researchers are actively discussing a staggering new service: the "$1,000 genome." A small desktop device in every laboratory or doctor's office would permit the rapid sequencing of anyone's entire sequence of DNA letters. With machines like this, researchers will be able to find the specific patterns of gene variations that contribute to complex genetic traits and disorders. For individual patients, they can use these findings to predict someone's likelihood of disease and better understand the many nondisease traits shaped by our genes, from hair and skin color to height, susceptibility to depression, and even IQ. As the *New York Times* science journalist Nicholas Wade observes, "When genomes can be decoded for $1,000, a baby may arrive home like a new computer, with its complete genetic operating instructions on a DVD."[3] It follows that access to the same information before a child is conceived or born will let us select desired traits for that child and, eventually, introduce new traits that neither parent possesses.

At the same time that genomic science and reprogenetic medicine are rapidly advancing, the United States is in an era of biomedical and bioethical retreat. Appealing to the conservative religious base, President George W. Bush has all but shut down aid for reproductive health initiatives around the world and crippled federal support for stem cell research in this country. His Council on Bioethics, under the direction of the conservative bioethicist Leon Kass, issued a series of reports largely seconding the president's views and questioning whether biomedicine had not already gone too far. The council's 2003 report, *Beyond Therapy,* took particular aim at efforts to use pharmaceutical, genetic, and other means to extend the human life span or enhance our children's genetic capabilities.[4] In the name of a "more natural bioethics," some council members and staff have spoken out against "artificial" biomedical interventions designed to prevent birth defects or help children born with grave diseases to live full lives.[5] In writings, Kass has repeatedly suggested that the current human life span is ideal and that the human genome should not be tampered with.

This retreat is not limited to those associated with the Bush administration. Daniel Callahan, one of the founders of modern bioethics, has repeatedly asked whether medicine has not gone too far in trying to ward off death. In books with titles like *Setting Limits, What Kind of Life,* and *The Troubled Dream of Life,* Callahan has argued for a transformation of our health-care system from one that seeks to conquer disease and extend life to one that aims at helping people accept and cope with the failings of the body.[6] The popular writer Bill McKibben has added his voice to this chorus of concern with a book entitled *Enough.*[7]

I, too, am a bioethicist, but I disagree profoundly with this conservative direction. I am deeply committed to progress in biomedical, reproductive, and genetic research. Over a career as a university-based scholar, National Institutes of Health (NIH) administrator, and advisor to private-sector biotechnology researchers, I have dedicated myself to supporting the development of new technologies aimed at assisting people faced with infertility, helping us understand and prevent the causes of birth defects, and using human embryo research to develop new approaches for tissue regeneration and organ replacement. I also believe that we should begin considering deliberate interventions in our own and our children's genetic makeup—to both prevent disease and enhance human life.

In this period of retreat, I want to draw attention to the impending revolution in genetic technology that will allow us to select or modify our children's genetic inheritance. I believe that the issue of gene selection and modification will dominate bioethics in the decades to come and emerge as a major focus of debate, dividing those opposed to biomedical advances from those committed to them. The forces driving our disagreements about this issue are not only ethical but also religious. At the center of our debates, we find very different views of God's sovereignty, the limits of human powers, and our appropriate responses to technological advances. These disagreements already fuel biomedical controversies at the beginning and end of life. In the years ahead, human gene selection and modification will become a leading focal point.

Genetic interventions raise many of our fears about biomedicine. Unlike stem cell research, which pits liberals against conservatives, left against right, genetic self-modification evokes opposition all across the political and cultural spectrum. Some liberal thinkers object to the expenditure of scarce research funds on what they regard as elitist genetic research. Others fear an emerging division between genetic haves and genetic have-nots. Some feminists who already see assisted reproductive technologies as a male usurpation of female reproductive powers join others opposed to the use of genetic testing for sex selection. The opposition includes environmental activists who believe that respect for untampered nature should extend to the human genome. In the United Nations, these constituencies have joined forces to help produce a UNESCO declaration calling the human genome the "heritage of humanity."[8]

A glance at the work of some of our leading writers and filmmakers shows how broad this coalition of opposition is. In the novel *Oryx and Crake* (2003), the feminist writer Margaret Atwood depicts a not-too-distant future where bioengineering has produced an apocalypse. Genetically altered viruses have killed off most of the human population. Aggressive, genetically modified, superintelligent animal predators, from dogs to pigs, roam the planet. Only a handful of human beings remain, but gene modifications have made them too innocent and vulnerable to survive. Atwood's novel criticizes our excessive love of science and our environmental intrusiveness. Her fear is that we can never retain full control of our creations.[9]

The same critical sentiment against human genetic self-modification is evident in Andrew Niccol's 1997 film *Gattaca*. The film was not a commercial success, but it lives on in bioethics classrooms around the country as the epitome of what is bad about human gene interventions. *Gattaca* begins with two acts of conception. The first takes place in the near future in the back seat of a Buick and produces Vincent, the love child of his young parents, Marie and Antonio. At birth, Vincent is diagnosed with a host of genetic maladies, including a heart problem that is predicted to end his life by age thirty.

The second conception occurs in the glistening laboratory offices of the Eighth Day Genetic Center and produces another son, Anton, one "worthy of his father's name." A counselor guides Marie and Antonio as they choose one of the embryos produced from their eggs and sperm through in vitro fertilization. He informs them of the modest genetic changes he has made in the embryos, including the eradication of genes for "prejudicial conditions" like premature baldness, myopia, alcoholism and addictive susceptibility, domestic violence, and obesity. When Marie and Antonio appear troubled by so much gene meddling, the counselor calms them, saying "You want to give your child the best possible start. . . . Keep in mind, this child is still you, simply the best of you. You could conceive naturally a thousand times and never get such a result." In a scene cut from the final version of the movie but available on the DVD, the counselor's speech to Marie and Antonio continues as he offers them a range of options. "For a little extra money," he says, "I could also attempt to insert [DNA] sequences associated with enhanced mathematical or musical ability." The couple decline the offer only when they learn that the cost is prohibitive.

Gattaca is meant to be a warning. Genetic manipulation leads to a nightmarish society obsessed with genetic perfection and disfigured by genetic discrimination. Vincent and other "nonengineered" people are scorned as "de-gene-erates" or "in-valids." They're consigned to the lowest rungs of society and shunned as mates. In the end, Vincent triumphs. Through hard work and determination, and with the help of other marginal people who evade the system, he achieves his dream of becoming an astronaut, proving that "there is no gene for the human spirit." But like so much science fiction dealing with genetics, the take-home lesson about human gene modification is wholly negative. In the film the deliberate manipulation of human genes threatens everything we hold dear: human individuality, freedom, justice, and love.

My aim in this book is to challenge the negative views that underlie works like *Oryx and Crake* and *Gattaca.* I do not intend to show that these visions are mistaken. No one can predict where our growing powers of genetic control will lead. The concerns voiced by artists, bioethicists, and others about genetics gone wrong are a healthy warning. In addition, the horrible eugenic abuses perpetrated by the Nazis that culminated in the Holocaust indicate how easily genetics can become an excuse for evil. Still, I believe that increased genetic control lies in our future and will make that future better. We will begin with gene selection aimed at reducing the likelihood that a child will be born with a genetic disease, and eventually include changes designed to permanently eliminate serious disorders like cystic fibrosis and sickle cell disease from a family line. Beyond this, gene modification will encompass the first hesitant steps to *improve* the genetic endowment of our children so they can flourish in new ways. This may include increased natural resistance to diseases like AIDS and cancer or to problems like diabetes or obesity. Somewhere down the line, we will see the emergence of what I call "cosmetico-genomics," as parents strive to give their children more attractive physical features, including normal height, good teeth, clear complexions, and pleasing faces. In the more distant future, we may see cognitive and neurological enhancements, ranging from reduced susceptibility to dyslexia, learning disorders, and depression to improved memory and enhanced IQ.

Some of these possibilities send chills down our spine because they raise so many scientific and ethical questions. What about mistakes and errors? Won't interventions carry unknown risks for the children themselves? Should we alter children's genes without their consent? What will these interventions do to the parent-child relationship? Will overzealous parents impose their vision of perfection on a child, creating psychological problems whether or not the child fulfills the parents' expectations? And what about social justice? Affluent parents have always been able to provide their offspring with advantages that enhance a child's chances of success. But now, in addition to better educational opportunities, a financial head start, and a network of well-placed friends, they could also confer "better genes." Will gene enhancement intensify and perpetuate the divide between society's winners and losers?

In the chapters ahead we will see that there are serious reasons for concern, but also some reasons to think that we will be able to adjust to (and even flourish in) a world where gene modification takes place. Above all, it is important to recognize how hard it is for us even to think about such new possibilities. Never before in history have we had anything like this ability to shape the biological inheritance of our children. Over millennia, we have become so accustomed to accepting our offspring as they are that the very idea of choosing their characteristics seems blasphemous and opposed to the way one human being should relate to another. These feelings and thoughts may be right. It could well be that the ability to choose another person's characteristics is a power that no human being should possess.

But our approach to this new power could merely reflect what social scientists call "status quo bias." Social science research has repeatedly shown that human beings resist change, even when there is no good reason to do so. In one study, researchers reported an episode that happened in Germany some years ago. The government found it necessary to develop an open pit coal mine, requiring the leveling of a village above the site, and offered to relocate the village to a similar valley nearby. Specialists developed planning options. Since the older village had evolved higgledy-piggledy over many centuries and was difficult to access by car, most of the new plans had obvious advantages. But in the end, the townspeople chose their familiar if inconvenient layout.[10]

The results can be explained partly by the emotional attachments people form to what they know, but many other studies in finance and economics, where no such attachment is evident, point to the same result: people tend to resist change and favor the status quo. When faced with change, they greatly overestimate the advantages of the familiar and conjure up reasons for disliking and fearing innovations.

Where human reproduction is concerned, to what extent are we like the villagers, favoring dysfunctional patterns just because we have grown used to them? It is always easy to identify the benefits of existing patterns: that they are there and seem to work is an argument in their favor. Every proposed change invites suspicion. But what we miss in this approach to novelty is a balanced and reasoned assessment of both the present and the future. What are the negatives, as well as the positives, in our present arrangements? In what ways is the present dysfunctional? Can we see ourselves growing accustomed to new patterns and even coming to regard them as better than present ones?

Imagining the future in terms of its benefits and losses is harder than criticizing the present. How do we begin to think about a world we have never seen, where people's relationships, capabilities, and values might be different from our own? How do we overcome the inclination to regard every change as a step into an abyss? One answer is to draw on some of the best works by writers of fiction who have extrapolated their perceptions of human life into a changed future. These writers' efforts are imperfect because, like us, they live in the present. The reach of their imaginations is constrained by known realities, and they are no less subject to status quo bias than the rest of us. It is not accidental that so much literature about human gene modification shares the nightmare character of *Oryx and Crake* or *Gattaca*.

The negative visions have much to teach us, but sometimes a writer is able to envision a world where human beings have successfully adapted to changed realities. In the chapters ahead, I will not neglect the negatives, but I will also draw on fictional works that offer a more positive vision of the role of gene modification in our future. For example, in the opening chapter, "Creating the Superathlete," I use a recent novel to ask whether the present "natural" state of athletic competition is as fair as it would be if gene doping leveled the playing field. Later on, I return to this question in a much broader context when I ask whether gene enhancements pose a threat to our ideals of a more just society. Will our skills at gene modification lead to a self-perpetuating "genobility," a genetic upper class so biologically different from the genetic "underclass" that the two groups no longer interbreed? Or will it foster our dreams of a more just society? In the novels of the writer Nancy Kress we find answers to these questions.

After examining the new capabilities that genomic science is bringing us, I again use a work of fiction to explore the safety of genetic interventions. Greg Bear's troubling short story "Sisters" reminds us how the most well-intentioned schemes can go awry. The writings of Nancy Kress and Greg Bear also raise questions about the impact of genetic choice on relationships within the family. Will gene enhancement technology inevitably deform the unconditional love that parents are expected to offer their children? Will it produce generations of troubled youngsters, plagued by self-doubt and anger at their parents and possibly exhibiting new forms of pride and arrogance? Or will it merely continue what parents have always tried to do: give their children the best tools for surmounting life's challenges?

More than we realize, our attitudes toward human gene modification are shaped by our religious traditions. Some of the opposition is overt, as when theologians warn us against the dangers of "playing God." Some is subtler. In a world where God is viewed as bringing nature into being, religion can offer powerful support to the status quo. Exploring religious views of gene modification, I look at the ways God has been invoked to caution against tampering with our genes. I also follow the African-American novelist Octavia Butler on a spiritual journey into a world beyond the human genome.

Discussions of human gene enhancement and gene modification tend to divide people into two camps: those bitterly opposed and those overwhelmingly favorable. Many of the opponents are thinkers who hold conservative religious beliefs. At the opposite extreme are the gene enthusiasts, who look forward to every new possibility as a sign of human creativity. Some of the enthusiasts are scientists, like the DNA co-discoverer James Watson and UCLA's Greg Stock.[11] Others identify themselves as "transhumanists," who look forward to a day when we will mold our biology as readily as we do plastic or steel.

I am in neither of these camps. I share some of the concerns of the conservatives. Human gene modification has the potential for great personal and social harm if it is introduced carelessly and allowed to proceed untended. But I believe, with near certainty, that sooner or later we will begin to modify our genes and that we will survive doing so. I have chosen the title *Babies by Design,* rather than the scornful alternative *Designer Babies,* because I believe that we are capable of bringing intelligence—"design" in the best sense of the word—to our reproductive lives. I am sure that we will make mistakes, most of which I hope we can correct and learn to avoid repeating. I also believe that eventually we will grow accustomed to a world where human beings make themselves physically and mentally better than they are today. Genetic science has opened our biology up to self-construction and directed evolution. We will certainly try to bring our biology under our control as we have done with so much of nature.

To put this view in perspective, consider how far we have come *without* genetic self-modification. Two centuries ago, the average human being died at about forty years of age; many people suffered from the effects of serious ailments like polio or tuberculosis, and others bore obvious deformities, from disfiguring skin lesions to clubfoot or cleft palate. Modern medicine has changed all this, extending life spans and eliminating many physical and cosmetic problems. According to Robert W. Fogel, a University of Chicago researcher on economic and population trends, new studies show that many chronic ailments, like heart disease, lung disease, and arthritis, are occurring, on average, ten to twenty-five years later than they used to. There is also less disability among older people today. Thanks to better nutrition, disease prevention, and treatments, says Fogel, humans in the industrialized world have undergone "a form of evolution that is unique not only to humankind, but unique among the 7,000 or so generations of humans who have ever inhabited the earth."[12] In the future, genetics will help us continue this evolutionary trajectory. Our descendants will live longer and be more vigorous. With the help of genetics, they may even do away with some of the newly introduced scourges of modern life, like the current epidemics of obesity and diabetes.

Behind much of our opposition to directed human evolution lies a subtle form of status quo bias: the belief that the human genome in its present form represents the highest expression of human biological possibility. While those who hold this view imagine cultural changes in the future that may take human civilization to higher levels, they generally believe that, at least in terms of genes, our biological nature has advanced as far as it can go. But this view is mistaken. The human genome is by no means a stable and finished phenomenon. Genomic research provides evidence that the human genome has continued to evolve rapidly beyond the appearance of anatomically modern human beings about 200,000 years ago.[13] Some of the change involves disease resistance. For example, when human beings started clearing African forests about 10,000 years ago they created an environment for malaria-bearing mosquitoes. In short order, the disease spread, and natural selection responded by leading to the proliferation of human gene mutations that enhanced people's malaria resistance. In Europe, it appears that a similar process worked in response to smallpox epidemics as recently as 1,300 years ago to alter the gene for a receptor that allows viruses to enter our cells. This change spread to only a small percentage of Europeans, but by an irony of history, the same change also confers resistance to HIV/AIDS.[14] As the AIDS pandemic continues throughout the world, will we see an acceleration of this ongoing genetic change?

Not all the changes involve disease resistance. About five or six thousand years ago, human populations in northwestern Europe became heavily dependent for survival on the whole milk derived from domesticated cattle. Throughout history, most human beings stopped drinking milk soon after being weaned from their mothers, and the genes that permitted them to digest milk sugar (lactose) were turned off. Around the world in cultures that are not milk-dependent the turning off of these genes still produces adult lactose intolerance. But in northern Europe (and in some cattle-raising regions of Africa), the opposite occurred: people with mutations in lactose-digesting genes survived and those that could not process this essential food died. Today, in most of Europe and America, we take it for granted that a glass of milk a day is good for you, but that is only because we are part of a newly evolved type of humanity.[15]

In September 2005 the prestigious journal *Science* carried two articles that took the recent nature of human evolution and its significance to new levels.[16] The articles detailed finding by teams led by Bruce Lahn, a human geneticist at the University of Chicago. One reported evidence of mutations that occurred about 37,000 years ago in a gene known as Microcephalin. This gene controls the number of neurons produced during fetal development. Individuals with seriously impaired versions of it suffer from microcephaly, a birth defect characterized by severe reductions in brain size coupled with mental retardation. But the mutations found by Lahn's team seem to have had the opposite effect. By increasing brain size or function, they conferred a selective advantage on those possessing them, and in a relatively short time the mutations were widely prevalent in European, Middle Eastern, and Asian populations. The second study found an even more recent mutation in another gene regulatory of brain size known as ASPM. This occurred about 5,800 years ago and spread rapidly through populations in Europe, the Middle East, and South Asia.

These two articles raise many questions. Are these genetic changes related to historical and cultural shifts? The report on the Microcephalin gene, for example, points out that the dating for the novel mutation corresponds to the start of what archaeologists call the "upper Paleolithic revolution" which was marked by the appearance of striking new uses of symbolism, including statuettes and cave paintings found at many sites in Europe. The time of the ASPM change roughly correlates with the rise of cities and spread of written language in the Middle East. Although neither article suggests causation, together they raise the possibility that these cultural changes may have been initiated or accelerated by changes at the genetic level. Much more controversial is the prevalence of these mutations in Europe, the Middle East, Asia, and the Americas but not in Africa. Lahn himself has tried to point out that other brain-enhancing mutations not examined in these studies may have arisen there independently.[17] Other studies support this, showing evidence of recent genomic changes under selective pressure in sub-Saharan African populations.[18] Nevertheless, the potential abuses of this information are clear, and some racist Web sites have already used it to spread their message of hate.[19]

Like all science, Lahn's teams' findings will remain tentative and suggestive until they are replicated and their implications are better understood.[20] But when combined with the other less controversial discoveries of the sweeping genetic changes in human populations that I mentioned, they offer us a new way of viewing the human genome. This is not the stable, once-and-for-all achievement that many have tended to believe. And we are not necessarily at the pinnacle of biological creation, as we have often fancied ourselves to be. Just as our physiology and neural architecture may differ from our not-so-distant ancestors', so the biology of our descendants may lead them to look, act, and think differently than we do. But what nature accomplished in the past by means of natural selection, we may do by direction. Emerging genetic technology permits us to replace the destructive and wasteful process of natural selection with intelligence and design.

If we are to achieve a transformed but recognizably *human* future, we have to avoid the realization of our nightmares. The worlds of *Oryx and Crake* and *Gattaca* could happen. We have to find ways of monitoring and moderating our steps before they do. I conclude this book by exploring ways of reducing the risks, and I offer guidelines for the responsible introduction of reproductive innovations.

My aim is to start us thinking now about developments that will someday, I believe, be a major focus of our social and ethical debates, but this technology is only in the earliest stages of development. The genetic enhancement of human traits and capacities will not be possible for a long time, and the most complex traits, like intelligence and temperament, may never be fully subject to control.[21] The philosopher Erik Parens cautions those who speak about bioethical matters to "strike a balance between responsibly contemplating theoretical possibilities like 'genetic enhancement' and responsibly conveying how little is currently understood about how such enhancements might be achieved."[22] I take this caution seriously, but I also believe that there are good reasons for starting to think about these possibilities. As I participate in workshops on genomics at the NIH and elsewhere, I am struck by how fast the science is moving. Possibilities that were almost unthinkable just a few years ago are now routinely being deployed in laboratories around the world. New microchip technologies, the unraveling of a host of mammalian genomes, including our own, and the ability to precisely insert genetic changes into organisms all work together to vastly augment our powers. Each development in one sector speeds breakthroughs in others.

Babies by design are in our future. As the molecular biologist Lee M. Silver observes, "Almost certainly, at some point a combination of scientific knowledge, technology, reduced risks, increased benefits, and societal acquiescence will cross a threshold, allowing human genetic engineering to proceed."[23] There are some who by prohibiting research would try to prevent this possibility from ever happening. Others want unfettered freedom to experiment. The challenge is to find a middle way. The time to start talking about this challenge is now.

CHAPTER 1

Creating the Superathlete

The first efforts to enhance human genetic traits may well take place in athletics. Athletes' willingness to accept risks in order to win makes them good candidates for gene modifications aimed at improving performance. The intense commitment that many parents now bring to a child's athletic career suggests that some parents may be willing to cooperate with reproductive specialists to increase their chances of bringing a future sports champion into the world.

In the past few years, molecular biologists have identified genes for a number of traits that directly affect athletic performance. The work of the University of Pennsylvania Medical School scientist H. Lee Sweeney is particularly striking.[1] Starting in the late 1990s, Sweeney and his collaborators sought to find new ways to combat the wasting effects of muscular dystrophy and aging. They knew that a naturally occurring substance, insulinlike growth factor I, or IGF-I, multiplies the number of cells responsible for the growth of muscle tissue. They also knew that if they injected IGF-I directly into muscle tissue, it would dissipate within a matter of hours. So Sweeney's team took a new tack. They isolated the sequence of DNA that manufactures IGF-I in our cells and, using recombinant gene techniques, inserted it into a population of harmless viruses that readily infect muscle tissues. They then exposed mice to the virus, hoping that the viruses would insert some of their genetic instructions for making IGF-I into the mice's DNA. Because these instructions reside in the genes, they'll produce IGF-I as long as the cells remain alive.

The results were remarkable. After injecting this altered virus into young mice, Sweeney's team found that the overall size of the muscles and the rate at which they grew were 15 to 30 percent greater than normal, even though the mice were sedentary. When the scientists injected the virus-gene combination into the muscles of middle-aged mice and allowed them to reach old age, their muscles did not become any weaker with age.

To further evaluate this approach, a colleague of Sweeney's at Harvard University, Nadia Rosenthal, altered the gene sequence in early-stage mouse embryos to create a line of mice that were engineered to overexpress IGF-I throughout their skeletal muscles. What resulted was a race of super strong rodents that Rosenthal dubbed "Schwarzenegger mice." The animals developed normally, except for having skeletal muscles that ranged from 20 to 50 percent larger than normal. As the mice grew older, their muscles retained the regenerative capacity typical of younger animals. Furthermore, their IGF-I levels were elevated only in the muscles, not in the bloodstream. This is important because high-circulating levels of IGF-I can cause cardiac problems and increase cancer risk.[2] Together, Sweeney's and Rosenthal's research proved that it is possible to genetically enhance muscle function and delay muscle degeneration, whether by gene therapy methods using viral vectors or by the direct modification of an embryo's genes.

Soon after the popular press picked up the story, Sweeney began receiving telephone calls from coaches and players around the country asking whether his research could be applied to human athletes. Some of the callers were interested in the finding that the injection of gene-altered viruses directly into the muscles could not be readily detected by standard blood-testing methods. In the minds of at least some athletes, Sweeney had found a way of enhancing performance that evaded the usual methods of drug detection. Although Sweeney and his colleagues began their research hoping to provide treatments or cures for disease and age-related muscle wasting, they had opened the door to gene doping.

Scientists have since explored several other approaches to enhancing muscle function. In addition to IGF-I, attention has focused on myostatin, a naturally occurring protein that seems to limit muscle growth throughout embryonic development and adult life. Mice that are genetically engineered to lack this antigrowth factor have considerably larger muscles than normal mice. Now scientists are looking at ways of inhibiting myostatin activity in muscle tissues as a means of repairing or improving muscle performance. Commenting on these findings, the physician-geneticist Philip R. Reilly observes, "My guess is that use of anti-myostatin drugs will surface as an issue in professional sports within the next few years."[3]

Investigators have also begun to explore some of the ninety other gene or chromosome variants that are associated with outstanding athletic abilities. For example, some athletes have genes that naturally endow them with a higher percentage of oxygen-carrying red blood cells. This is especially valuable in sports that involve extended exertion, like marathon running or professional bicycle racing. In 1998 the Tour de France was nearly cancelled after an assistant for the Festina team was caught with hundreds of vials of erythropoietin (EPO). The illicit use of erythropoietin by racing competitors to grow red blood cells shows how attractive this advantage can be, and makes the possibility of less detectable genetic means of improving oxygen transport of great interest. In a different direction, Australian researchers recently examined a gene called ACTN3 in a group of male and female sprinters. A small variation in this gene apparently gives rise to a higher proportion of the fast-twitch muscle fibers used in high-speed, short-distance events. The scientists found an unusually high frequency of this ACTN3 variant in these elite sprinters. In particular, more of the female athletes had two copies of the variant than would be expected in a randomly selected group.[4]

The possibility of gene manipulation has already drawn the attention of sports authorities. The International Olympic Committee's World Anti-Doping Agency (WADA) has sponsored several scientific summits to determine whether it is possible to develop means of detecting altered gene sequences in the body. At least one leading gene therapy researcher, Theodore Friedmann of the University of California, San Diego, believes it is. According to Friedmann, every biomedical intervention leaves its traces. Genes altered in leg muscles, for example, are likely to cause small but detectable patterns of altered protein expression in cells elsewhere in the body.[5] The challenge before the scientists working with anti-doping agencies is to develop the sensitive assays needed to detect these changes.

But why should they? What's wrong with gene doping, and why shouldn't we regard it as just another tool, like better food or innovative training techniques, to improve competitiveness? It is true that steroids and EPO carry serious risks for their users. A whole generation of East German women athletes was exposed to serious endocrine problems leading to the growth of body hair, other masculine attributes, and sterility when their coaches surreptitiously gave them high doses of steroids. Hundreds of those athletes have applied for post-injury compensation, and many more who were injured have remained silent out of shame.[6] In the late 1980s, before athletes figured out how to use the drug, about twenty Dutch and Belgian cyclists died when their EPO-thickened blood caused cardiac arrest. Some fear that unusually strong muscles might tear away ligaments or tendons during intense competition.[7] But since gene therapy can be aimed only at improving performance in peripheral muscles, there is good reason to think that it will have fewer harmful systemwide effects than steroids or EPO. Why, then, are there ethical objections?

Kevin Joseph's novel *The Champion Maker* (2005) offers us a fictional glimpse into the world of gene enhancement in sports. It gives us a way to imagine the opportunities and problems of this technology and anticipate some of the moral choices before us. It is a good place to start our journey into the world of gene modification and enhancement.

The hero of *The Champion Maker,* Stanley "Bullet" Gardner, is a brash teenager more interested in drinking beer and playing rock music with his garage band than pursuing high school studies, but his running abilities amaze his friends and attract the attention of Nick Vance, a writer for *Running Strong* magazine. Nick's own track and field career was terminated by the U.S. Olympic boycott after the Russian invasion of Afghanistan. That and the suicide of his young wife sent Nick into a two-decades-long downward spiral. In Bullet, Nick sees the chance for a great story and, possibly, his own redemption as coach and advisor.

Bullet wants to compete in the Olympic qualifying trials for both the 100-meter and the 1,500-meter events. Realizing that these require very different skills, Nick counsels against the dual effort. He relents when he witnesses Bullet's exceptional abilities. After intensive training, Bullet wins both races.

Then, just weeks before the Olympics, everything goes wrong. The U.S. Doping Prevention Agency and World Track Federation accuse Bullet of using gene doping to enhance his abilities. Tests reveal that he has an exceptionally high level of hemoglobin in his blood, enhancing his metabolism of oxygen. His muscle fibers show signs of a gene variant that allows rapid switching between long-distance and sprinting capabilities. Bullet denies any involvement with drugs or gene therapies.

Aided by Tori Lee, a Washington sports lawyer, Nick solves the mystery. He learns that Bullet is one of many young people produced during the Cold War by a covert Defense Department research program known as Double Helix. Directed by an ex-Nazi geneticist, Lothar Schenck, the program aimed to produce superhumans for war and international competition. Using assisted reproductive technologies to modify early-stage embryos, it produced 620 children with genetically enhanced intelligence, longevity, strength, speed, endurance, and agility. Shortly after their birth to surrogate mothers, the children were placed for adoption with parents who were sworn to secrecy.

Nick, Bullet, and Tori fight to prove Bullet's innocence. The novel reaches its climax in the courtroom of U.S. District Court Judge Harold Thornberg, where Tori fights the doping charges. She asserts that it is impossible to separate the role played by Bullet's genes in his success on the track from his determination and enormous training efforts. "For every Stanley Gardner out there setting records, your honor," she says, "there are most likely scores of lazy, uninspired human beings with the same natural gifts but without the dedication necessary to be a champion."[8]

But the heart of Tori's plea is her claim that Bullet's genetic status does not violate the anti-doping code's definition of gene or cell doping as the "non-therapeutic use of genes, genetic elements and/or cells that have the capacity to enhance athletic performance." Although she acknowledges the argument of the chief attorney of the World Track Federation (WTF) that the prohibition seems to apply to any kind of gene manipulation, she contends that it does not cover preconception manipulations. Its real purpose, she insists, is to prevent athletes from abusing gene therapy, not to check the participation of so-called designer babies in sports—it is intended to avoid the imminent menace, not the future threat. She adds that a rule barring someone from athletic competition because of genetic attributes that they inherited but did nothing to produce would exclude from competition such world-class athletes as Tiger Woods, Lance Armstrong, and Michael Phelps.

Judge Thornberg is moved by Tori's arguments, but he worries that allowing unfettered genetic engineering "will make a sham of international athletics by rewarding those countries with the resources and know-how to build the strongest and fastest athletes."[9]

In the end, events outside the courtroom, not the judge's opinion, settle the matter. Lothar Schenck resorts to violence to protect his program, forcing the Department of Defense to intervene. Schenck is removed from command of the Double Helix program, and the WTF is urged to withdraw its opposition to Bullet's participation in the upcoming events as long as he retires from competition after the Olympics. The novel ends happily with Bullet winning two gold medals, and Nick marrying Tori and opening a runners' training camp.

The Champion Maker may never enter the ranks of great literature. Like many novels that use our anxieties about emerging scientific developments to drive the plot, it serves as an engaging diversion for the beach or airplane. But the very appearance of this book is important. It signals a growing awareness of gene doping and suggests that, along with anabolic steroids and human growth hormone to build muscle and EPO to multiply red blood cells, genetic interventions may soon be used to boost athletic performance. It adds to the debate about what kinds of genetic modifications we are prepared to permit in athletics—or human life generally. That Tori Lee's courtroom defense of Bullet is not entirely convincing even to Bullet's coach and mentor, Nick Vance, tells us that our moral intuitions about the use of genetics to enhance athletic abilities are uncertain. On the one hand, we worry that the use of genes to produce a race of "über-athletes" could undermine our very notion of sports competition.[10] But we also know that genes play a major role in athletic performance, and we wonder where we should draw the line between what is permissible and what is off-limits.

Safety is not an issue in *The Champion Maker.* When arguing before the judge, not even the lawyer for the WTF is willing to claim that Bullet has been harmed by the techniques that have made him so extraordinary. Is it fairness, then, that stirs our concern? Certainly it seems reasonable to oppose a concealed innovation that gives one competitor an edge over others. We tend to think that athletes should compete on a level playing field and that artificial performance enhancements distort that field. But even a moment's reflection shows that athletes do all sorts of things to give themselves an advantage, such as improved nutritional programs and intensive training efforts (often designed and assisted by sports medicine professionals).[11] Although anti-doping authorities have banned EPO, many athletes use high-altitude (hypobaric) training regimens and devices to increase the oxygen-carrying capacity of their blood. "Live high and compete low" has become a mantra for some professional sports trainers. Athletes seeking an edge have introduced such equipment improvements as graphite tennis rackets, fiberglass vaulting poles, and Speedo "Fastskin" neoprene swimsuits. After some initial debate, these have all now have become commonplace in Olympic and professional competition. Tentative efforts by the World Anti-Doping Agency to rein in high-altitude training tents have met with a barrage of criticism from athletes and trainers who have come to rely on these aids.[12]

So not every use of "artificial" technical improvements is unfair. We allow athletes many ways of seeking competitive advantage that go beyond hard work and grit. These start as efforts to tilt the playing field in the individual athlete's favor, but others soon learn to take advantage of them. Only some of these efforts end up being banned as unfair. This usually happens because we have concluded on other grounds that this particular technical improvement should not be allowed. We prohibit steroids not because using them is unfair but because they are physically dangerous and because we do not want to create an environment where everyone is pressured to use them. Once the ban is in effect, we label their use as unfair and regard anyone who does so as a cheat. Fairness, in other words, is not so much an argument against technical innovation as it is the conclusion of an argument based on other considerations. As Andy Miah puts it in his thorough study of gene doping in sports, "The idea that cheating or rule breaking is determined solely by what is outside the rules begs the question as to what ought to constitute the rules in the first place."[13] But this brings us back to the question of why we want to label gene doping as unfair.

Regarding gene doping as unfair becomes even more puzzling when we consider how important genetic differences are in athletics. Consider the case of Eero Maentyranta. For decades after he won two gold medals in cross-country skiing events in the 1964 Winter Games at Innsbruck, he was dogged by charges of blood doping—adding red blood cells before the race to increase his oxygen and stamina. The practice was common among cheaters during that era. Tests showed that Maentyranta had 15 percent more blood cells than normal, but with no evidence of doping the controversy just smoldered, and Maentyranta ended up with seven medals won in three Olympics. When genetic testing became available, Finnish researchers found that Maentyranta and other members of his family carry a rare gene mutation that naturally produced the EPO hormone and loaded his blood with 25–50 percent more red cells than the average person has.[14] Maentyranta was a natural genetic wonder, and the blood doping charges were forgotten. But if he had injected himself with blood cells or a gene-carrying virus to achieve the same result, he would have been branded a cheat and stripped of his medals.

Or consider the case of my own favorite athlete, Lance Armstrong. Over the years, Armstrong has been dogged by accusations of drug use and blood doping. Whatever the truth of these charges, he is a physical prodigy. Sports physicians have studied few competitors as intensely as this seven-times Tour de France winner. Armstrong's heart is almost a third larger than the average man's. During the rare moments when he is at rest, his heart beats about thirty-two times a minute—slow enough to cause a doctor who knew nothing about him to call an ambulance. But when Armstrong is racing, his heart rate can rise above two hundred beats a minute.[15] His other attributes are equally amazing. His VO_2 max—the maximum amount of oxygen his lungs can take in, an important measurement for an endurance athlete—is extremely high. He has long femur bones in his legs, giving him extra leverage on the pedals. And he has an extremely high proportion of slow-twitch muscle fibers, the type used by endurance athletes.

Some of Armstrong's physical attributes have been honed by years of training. Intensive efforts, for example, have shifted his balance of slow muscle fibers from good to excellent. According to Phillip Sparling, a professor of applied physiology at the Georgia Institute of Technology in Atlanta, Armstrong "is on top of the cycling world because of the combination and interaction of his genetic endowment, years of incredible training, competitive experience, and obsessive drive to achieve and persevere."[16] And yet even Armstrong's drive may have a genetic component. A person's nervous system partly determines how well he or she can train and how successfully he or she makes adjustments and improves technique. Lance's mother, Linda Armstrong Kelly, showed similar drive and determination when, starting as a seventeen-year-old single parent with only a high school education, she raised her boy to athletic stardom and herself to career success.[17]

We do not have to solve the ancient mystery of nature versus nurture to see that genes play an important role in athletic achievement. They may only be the foundation on which performance is built, but in many sports that foundation is needed to even compete. Although only a few tall people become professional basketball superstars, almost no one lacking the genes for height can be a serious basketball competitor today. The renowned Swedish exercise physiologist Per-Olof Åstrand overstates matters, but not by much, when he says, "The most important thing an aspiring athlete can do is to choose the right parents."[18]

If so much of athletic accomplishment is a result of our genes, and if great athletes start life with a better-than-average genetic endowment, why is it unfair for someone to try to level the playing field in their own—or their children's—favor by gene doping? Why is the natural lottery of genetic inequality fair, but measures to correct that inequality are not?

Let me speak personally here. Like many people, I am athletically challenged. In grade school, I could not catch or hit baseballs. I was good at other things: running, even throwing balls. But when I came to bat, my teammates groaned. In gym, with other misfits, I was among the last to be chosen for anyone's side.

When I became an adult, in the era of video games, I began to understand my problem. I am terrific at games that require me to move, aim, or steer: firing a rifle or ray gun, steering cars and planes. But I cannot connect with incoming objects and often miss them, scoring very poorly in games involving this ability. This is probably the result of some odd wiring in my brain. For a professor of bioethics not being able to connect is not a big problem, but in the world of American school athletics, where just about every game involves catching or hitting balls, this failure was enough to stifle my athletic ambitions.

In athletics, life's not fair. Would it make sense, then, to use drugs or genetics to level the playing field to accommodate those who lack athletic abilities? Julian Savulescu, a British-Australian bioethicist, thinks so. Writing with two colleagues, he argues that "sport discriminates against the genetically unfit. Sport is the province of the genetic elite (or freak)." Savulescu believes that we should allow moderate performance-enhancing drugs in sports. His argument applies to gene doping as well. "The result," he says, "will be that the winner is not the person who was born with the best genetic potential to be the strongest. Sport would be less of a genetic lottery. The winner will be the person with a combination of the genetic potential, training, psychology and judgment. Olympic performance would be the result of human creativity and choice, not a very expensive horse race."[19] A German professional cyclist reinforces Savulescu's point when he says, "EPO lifts all pro cyclists up to the same level, and the only difference now is what you can accomplish by training."[20]

Even with my athletic deficits, I do not agree with Savulescu. Despite everything I have said about the unfairness of the genetic lottery and the poverty of arguments based on fairness alone, there are good reasons for wanting to limit the amount of technical—and biotechnological—intervention we permit in athletics. For starters, there is the safety issue. Any biochemical intervention in the human body carries risks, many of them unknown. Since athletic competition is so intense, and the quest for gold so powerful, athletes are tempted to use to excess anything that gives them an edge. Once we open the door to drug-enhanced performance, it will be very hard to close. Furthermore, no one will be better off if we permit drug or gene use. The goal in sports is always primarily competitive or "positional" advantage: doing better than others. Once everyone can use drugs or gene therapies to boost performance, no single person is better off. In ranking terms, each competitor is back where he or she started, although everyone's risk is heightened.

Finally, there is the impact of such interventions on the nature of the sport itself. Do we want to move the focus of athletic competition from the playing field to the laboratory? Should Olympic gold be determined not by an athlete's native abilities and determination but by the quality of the scientists he or she hires? Before you reply "Of course not!" let me observe, in my more cynical voice, that for many people the attraction of sports is not just the chance to watch great athletes demonstrate their abilities. The World Anti-Doping Agency protects the "spirit of sport," which it defines in terms of specified values: ethics, fair play and honesty, excellence in performance, character and education, fun and joy, teamwork, dedication and commitment, respect for rules and laws, respect for self and other participants, courage, and community and solidarity.[21] But high school coaches eager to win games have sidelined me often enough that I tend to regard this as romantic claptrap. The booming hometown attendance records for Giants' slugger Barry Bonds, despite allegations of his steroid use, show that many people think sport is primarily about winning and setting records. As one fan observed when Bonds surpassed Babe Ruth's number of career home runs, "Steroids or no steroids, the man pulled it off."[22] This tells us that the meaning, or "spirit," of sport is surely relevant to our consideration of what kinds of enhancements we are going to permit, but it is only one factor in our decision making. Sheer excellence in performance, however achieved, is another. If you doubt this, recall that professional athletes, once barred from the Olympics, now routinely participate. The spirit of sport often takes second place to people's interest in witnessing the most exciting competition and seeing the strongest people compete.

The issues raised by human genetic self-modification and gene enhancement include many that I have not mentioned at all, including the use of gene technology to design and deliberately bring into being a child with enhanced capabilities. Bullet is one such child, but his story is hardly the best example of what could happen. His designer was a crazed ex-Nazi with no interest in his engineered offspring's well-being. Loving parents raise different issues. Parents who are passionate about some type of excellence, whether it is connected with running, baseball, or music, might be tempted to use gene modifications to give their child an improved start in life. This raises further questions. To what extent will this technology encourage parents to use children as means to their own goals? How will children respond to the knowledge that they were brought into being for someone else's purposes? In a short while, we will address these questions directly. But for now, they are only at the margin of our vision.

Is it ever permissible to artificially enhance the talents that people bring to competition? The answer is not as clear as it might seem. Although gene doping appears to be nothing more than cheating, it stands somewhere along a spectrum of interventions ranging from those that are clearly allowable, like high-altitude training and better nutrition, to those that offend us, like steroid use. Furthermore, unlike other interventions, genes are at the very heart of athletic performance, and they are not distributed fairly. This raises the question of why we would ban gene doping or prenatal genetic enhancement when we celebrate nature's unequal bestowal of athletic talent. It also raises the question of whether gene therapies might be a useful way of rectifying inequalities of birth in order to bring about a world where thousands of average competitors, including athletically challenged people like me, have the chance to become a Lance Armstrong or Eero Maentyranta.

When we compare athletic enhancements with gene enhancements in general, things become even less clear. Sports enhancements aim primarily at positional advantage rather than absolute levels of achievement. An athlete hopes to excel over others in competition, but if everyone uses the enhancement, no one is better off (and all competitors may be harmed by the risky innovation). In much of human life, however, improved abilities provide an absolute benefit to their possessor and possibly to society. An airplane pilot with improved vision and quick reactions might avoid an accident. An Einstein-like mathematician or physicist could help us gain important insights into the nature of the universe. A neurosurgeon with exceptionally steady hands might save a patient's life. In ways that are not true in athletics, efforts to enhance certain human capabilities make sense, even if they involve some added risk to their possessors.

My aim so far has been to trouble our thinking. In chapter 6 we will look more closely at how gene enhancements might affect our notions of fairness and justice. Will enhancements give some an advantage in life's struggles and increase the gap between life's winners and losers? Or will they provide new ways of correcting inequalities, including those we now accept simply because they are natural?

Our thinking about such complex possibilities rarely hinges on a single factor. Take the safety issue. This is surely a major concern, but it is not the only matter that shapes our judgment. If safety in sports were the only consideration, we would not permit boxing or downhill ski racing. Instead, we balance safety against possible benefits and opportunities. Fairness, too, is an important consideration, but, as we have seen, judgments of fairness are usually the outcome of a many-sided decision process. Our thinking about what is right includes other factors that we tend to overlook. One is whether a technical enhancement, despite its artificiality, will increase the interest or value of the activity. Fiberglass vaulting poles and neoprene racing suits have passed this test, but the development of ultralight racing bicycles has been banned because it places too much emphasis on technical innovation. Another consideration is whether we can realistically prevent or discourage an enhancement's use. High-tech nutrition, including "carbo-loading," is an artificial enhancement, but we do not ban it because it is too hard to control what athletes eat. An emphasis on nutrition shifts the locus of achievement from athletes to their cooks, but since the intervention is harmless and hard to monitor, we ignore it.

We will see that a multifaceted decision process applies to all of our thinking about genetic enhancement. We cannot reject an enhancement just because it's "unnatural" or because it poses questions of fairness or justice. Instead, we have to balance safety, fairness, justice, controllability, and the effects on parents, children, and society. Sometimes this balancing will prompt us to ban a type of enhancement. At other times, it will lead us toward deliberate genetic changes as a part of the everyday repertoire of medicine.

CHAPTER 2

How Will We Do It?

Mario Capecchi has invented a technology that could change our world. Seated in his fifth-floor office in the Eccles Institute of Human Genetics on the University of Utah campus, with the Great Salt Lake and the Oquirrh mountain range shimmering in the window behind him, he reviews the ways that genome research is expanding our ability to understand and alter human genetic inheritance. Capecchi's voice rises with emotion when he considers whether we should go beyond genetic selection to actively modify the human genome. "We're not close enough to understanding the issues to make wise decisions or predict the outcomes," he insists.

Capecchi's caution is shared by many other molecular biologists working at the forefront of genomic science today. But unlike many of these scientists, he has discovered a tool likely to make human gene modification a reality. The tool is homologous recombination, a utilization of the cell's own gene repair mechanisms to make site-specific gene targeting and gene alteration possible. Capecchi's work in developing homologous recombination has earned him scientific prizes around the world. Yet in 1980, when he first submitted an application to fund experiments testing the feasibility of gene targeting in mammalian cells, NIH reviewers rejected his grant proposal, labeling it "not worthy of pursuit." Capecchi persevered. When he resubmitted his proposal four years later, the reviewers not only enthusiastically approved his grant but also extended an apology: "We are glad that you didn't follow our advice."[1]

To understand the revolutionary importance of Capecchi's discovery/invention, it helps to tell the story of some of the most innovative gene therapy research today. All around the world, scientists are seeking ways of changing aberrant gene sequences that cause disease. Take the rare genetic disorder known as X-linked Severe Combined Immune Deficiency (X-SCID). Think of this as a genetically inherited, as opposed to a virally induced, form of AIDS. Infants born with X-SCID carry misspellings in the sequence of DNA letters in a gene (known as IL2RG) that codes for the proteins that make up the interleukin-2 receptor. These receptors are needed for the proper functioning of key cells of the immune system. Because of X-SCID patients' DNA defect, they have very few T cells and lack the antibody and natural killer cells needed to fight infection. A generation ago, children with X-SCID died before the age of two; they survived longer only if they were kept in sterile environments, like David Vetter, the "boy in the bubble."[2] Today, most children with X-SCID are able to survive thanks to matched bone marrow transplants from close relatives. But bone marrow therapy fails for some of these children, and they face an inevitable downward course.

In 2000 a team of researchers at the Necker Hospital for Sick Children in Paris tried a new approach. Working with ten children with X-SCID who had not responded to bone marrow therapy, they removed a sample of blood cells from each child and exposed the cells to an otherwise harmless retrovirus that had been altered to carry a corrected DNA sequence for the IL2RG gene. Retroviruses are able to insert their DNA into the cells they infect. The genetically modified blood cells were then transfused back into each child and allowed to multiply. Within a short period of time, nine of the children were producing enough healthy immune cells to fight off infections. This was the first major success in gene therapy. Unfortunately, it was not a complete success. Within a year or two of the treatment, two of the ten children developed leukemia (a third later contracted the disease).[3] Research showed that in each case, the corrected IL2RG had misinserted itself into each child's genome, usually falling into a region known to activate a gene related to leukemia. It was as if I had mistyped this sentence and accidentally pushed a repeater key (xxxxxxxxxxxxxxxxxxxxx) that triggered cancer. Although this may have happened in only a few of the millions of transformed cells transfused back into each child, those cancer-prone cells had a tendency to proliferate and soon outnumbered normal cells. Fortunately, leukemia is usually treatable in children. To date, only one of the children in this experiment has died, and the others, who would probably have succumbed to X-SCID by now, are doing well. On balance, the Necker experiment remains a success.

This experience reveals a major problem in gene therapy research. Many of the delivery vehicles (or "vectors") used to carry corrected gene sequences into cells are nonspecific: they deposit their DNA payload randomly among the three billion paired nucleotide letters of the genome. Most of the time this is harmless and even beneficial. Like little freestanding protein factories, the inserted genes can function normally just about anywhere in the genome, remedying the targeted deficiency or curing the targeted disorder. But sometimes the genes or sequences misinsert and disrupt the functioning of normal genes. This is termed "insertional mutagenesis." As in the case of the X-SCID children in France, it can lead to an artificially induced form of genetic malfunction that may be as serious as or even more serious than the original disease.[4]

Against this background, we can better understand the revolutionary importance of Mario Capecchi's work. To find a more precise way of inserting genetic material into an organism's genome, Capecchi looked to a natural process that occurs when DNA becomes damaged in a cell. In such cases, the genetic machinery of the cell somehow has a way of identifying a misspelled sequence of DNA letters on one of the two paired strands of DNA in the double helix. It then cuts it out and replaces it with the correct version from the accompanying strand. (In the formation of sex cells, this same process of "homologous recombination" works to cut and paste large blocks of similar DNA from one strand to the other, allowing the organism to exchange similar traits from the maternal and paternal DNA strands of each chromosome.) To this day, it is not quite clear how the genetic machinery of homologous recombination works, but Capecchi was able to harness this process to permit direct, site-specific insertion of DNA sequences into the cells of selected organisms. Identifying a target DNA sequence in a bacterium, yeast, or mouse cell, he would prepare a similar sequence of his own and then use a variety of techniques to cause homologous recombination to occur. The result was a cell with the old DNA sequence snipped out, and the new one precisely inserted in its place. Apart from its opening and closing letters, this new sequence did not have to contain every DNA letter of the old one; substantial similarity (or "homology") was enough. Capecchi had thus invented a way of directly altering the DNA in cells without incurring the problem of insertional mutagenesis that besets most other forms of gene transfer technology.

Why didn't the French researchers use homologous recombination instead of retroviruses to insert the corrected sequence for the gene? The answer is that homologous recombination is very inefficient. If you expose a large population of cells in a petri dish to a replacement sequence, only a few of the cells (perhaps one in a million) will take up the new sequence. Retroviruses and the other viral vectors used in gene transfer are far more efficient, potentially infecting almost all the cells and changing the DNA in many of them. Viral vectors are not free of problems. In addition to insertional mutagenesis, they can cause severe immune reactions in some patients. In 1999 such a reaction killed an eighteen-year-old man, Jesse Gelsinger, in a gene therapy experiment at the University of Pennsylvania.[5] But the very low efficiency of homologous recombination makes even risky viral vectors an attractive alternative in gene therapy research.

In early January 2006, a few months after meeting with Mario Capecchi, I visited Theodore Friedmann in his office at the medical school of the University of San Diego. Friedmann is one of the leading scientists working on human gene therapy. He has served as chairman of the Recombinant DNA Advisory Committee (RAC), the NIH group that oversees and regulates all such research in this country. Because of his regulatory background and experience, Friedmann is even more cautious about the prospects for safe human gene modification than Capecchi. Shortly before he became chairman of the RAC, the Gelsinger tragedy occurred, and he wrote articles tracing the long road ahead for gene therapy research.[6] Yet I had barely settled into my chair in Friedmann's crowded office when he volunteered the opinion that the science is moving so fast that things he regarded as impossible just a few years ago are already happening.

Friedmann was particularly impressed with reports about the work of Aaron Klug and his lab at Cambridge University. Their research aims at making sure that the machinery of homologous recombination finds its target. The key is a specially engineered protein called a "zinc finger nuclease" (ZFN) that can be tailor-made for any identifiable DNA sequence. ZFNs search out the desired DNA sequence and home in on it like a guided missile, increasing the efficiency of homologous recombination a thousandfold.[7] This is the first, but certainly not the last, breakthrough in a research direction that will someday give us the power to change DNA in living cells without fearing that we will be disturbing other features of the genome.

For two hours, Friedmann and I wended our way through the complexities of gene therapy research and the prospects of gene doping in athletics. In his role as a scientific advisor to WADA, the World Anti-Doping Agency, Friedmann has argued for the need to develop reliable assays for gene alterations if we hope to control the misuse of genetics in sports. As we ended our conversation, Friedmann reiterated his belief in the complexity of the human genome and his worries about the harmful effects of gene modifications. The science in some areas is moving faster than our ability to use it carefully, he said. But he acknowledged that the pace of developments surprises him.

Homologous recombination is not the only route to targeted gene modifications. Some researchers are tinkering with the idea of human artificial chromosomes (HACs) as a way of getting new DNA into the genome.[8] Chromosomes are the relatively isolated islands of DNA across which the three billion pairs of nucleotide letters of the genome are distributed. There are forty-six of them in the human genome. By adding a new island to this geography, scientists can insert new genes along with the upstream "promoter" regions that turn them on and off. Since the DNA on these new islands would not integrate into the normal forty-six chromosomes, HACs could, in theory, be introduced into embryos without fear of disrupting existing genes on the other chromosomes. Although artificial chromosomes have proved functional in bacteria and yeast cells, there are problems with this approach. Extra chromosomes in human beings are often associated with disease, and it is not clear that this would not also be true with HACs.[9] In addition, for successful reproduction, chromosomes must be matched with similar chromosomes from a sexual partner. Unlike with targeted gene changes in existing DNA, therefore, someone who received new genes via artificial chromosomes would not be able to transmit them to the next generation unless his or her mate had the same artificial chromosomes. In some cases, this one-generation limit on gene changes might be desirable, but in general the application of homologous recombination seems to be the best way of achieving gene modifications in the future.

The ability to change DNA in human cells is only one of the complex skills we will need to accomplish human gene modification. Another is the ability to understand what each part of the genome does. Before we modify genes, we must know what we want to change. Then we need the ability to inject modified gene sequences and cells into a living individual to achieve specific and long-lasting physical effects. Genomic science is making astonishing advances in all areas of the gene modification process.

In the past few years, researchers have made enormous progress in learning how DNA creates and shapes the human body, and the pace of learning is increasing exponentially. A major reason is the Human Genome Project. In 2003, the HGP made available on publicly accessible Web sites the entire sequence of nucleotide letters, the three billion As, Cs, Ts, and Gs that make up the human genome. Now scientists interested in finding a gene or understanding the meaning of a specific stretch of DNA do not have to waste months of laboratory time sequencing DNA on their own. Instead, a simple computer search locates targets of interest. Each new discovery of the meaning of parts of the genome is added to the list of annotations that accompanies the public sequence. Knowledge builds on knowledge to accelerate the pace of discovery.

Almost as important as the HGP map is the sequencing in 2002 and 2003, respectively, of drafts of the genomes of the mouse and the chimpanzee. Both mammals share a great deal of genetic sequence with us. On a letter-by-letter basis, the mouse's genome is identical to ours in nearly 85 percent of the parts of the sequence that code for genes, which are the functional units that make proteins and other key building blocks of the body; the chimpanzee, our closest evolutionary relative, shares perfect identity with 96 percent of all of our DNA and 98–99 percent identity in gene coding regions.[10]

Having the mouse genome in hand is particularly important for helping us understand just what a gene or DNA sequence does. The mouse, a small and fast-breeding laboratory animal, can be used to explore gene function and dysfunction. Over the past few decades, researchers have become experts in producing made-to-order mice with specific DNA sequences either "knocked in" or "knocked out." They begin with a population of undifferentiated stem cells culled from mouse embryos and apply gene modification techniques, such as homologous recombination or viral infection, to make a genetically altered or "mutant" cell. A factor conferring resistance to an antibiotic is attached to the inserted gene to identify the small number of stem cells in which the desired gene change occurs. When the cells are dosed with the antibiotic, only those cells carrying the resistance factor *and* the desired DNA sequence survive. These modified stem cells are then injected back into a few-days-old mouse embryo, and, as the embryo develops, the mutant cells proliferate alongside its normal cells. What results is a "chimeric" mouse, with both normal and mutant cells permeating its body (the term is drawn from the Chimera of Greek mythology, a blend of a lion, a goat, and a serpent). The final step is to mate this chimeric mouse with another like it. Since some modified cells find their way into eggs and sperm, two chimeric mice can produce an embryo with two copies of the mutant gene in all of its cells. If dark coat color genes are linked with the modification, the pure mutant mouse shows up in the laboratory when two mottled chimeric mice give birth to a dark brown pup.

Scientists call this a "transgenic" animal, one that has had foreign DNA stably integrated into its genome. Depending on the gene modification, some of the mouse embryos produced in this way are never born, because the modification interrupts normal development. This can be instructive in illustrating the importance of the gene sequence involved and the lethal implications of its modification. But even more useful are changes that slightly alter the physical or behavioral functioning of the resulting animal. A mutant "knockout" mouse lacking the key genetic information needed to form cell membrane components for conducting sodium ions, for example, will develop a condition very much like cystic fibrosis. Since the mouse and human genes in this region are very similar, this offers scientists a way of understanding the corresponding role of the analogous DNA misspelling in human beings, and it also yields a line of transgenic mice that can be used to test new drugs and therapies for treatment of the human disease. In addition to physiological changes, genetically influenced cognitive and behavioral phenomena can be studied. Using knockin-knockout technology, scientists have produced, among other things, lines of mice displaying the features of Alzheimer's and Parkinson's disease, lines resistant to morphine addiction, and lines displaying hyperactivity and increased male-to-male aggression.

Long-lived, large, and strong-willed animals like chimpanzees do not furnish as good a test bed for trying out gene modifications. The importance of the chimpanzee genome lies elsewhere—in its even greater similarity to the human one. A mouse shares perfect identity with 85 percent of our DNA in gene coding regions compared with the chimpanzee's 96 percent. By studying the subtle variations in sequence between chimpanzees and ourselves, scientists are gaining new insight into the specific genetic features that have made us human.[11] For example, an international team led by the evolutionary biologist Gregory Wray of Duke University recently found that the functioning of a gene that codes for the protein prodynorphin (PDYN) differs substantially in chimpanzees and humans. PDYN is a precursor to a number of endorphins (opiatelike molecules involved in learning, the experience of pain, and social attachment and bonding). Although chimpanzees and humans have the same PDYN gene, a promoter sequence just upstream from the gene's coding region is far more active in human beings. Similarly, scientists have found significant differences in the FOXP2 gene in the two species. This gene is associated with speech acquisition. Researchers studying a large British family, many of whose members have barely intelligible speech, found that affected family members have mutations in the FOXP2 sequence that make it more like the chimpanzee gene than the human one.[12]

Studying the chimpanzee genome more closely, therefore, may shed light on just what happened in evolution to form our species. Recent evidence suggests that our early human ancestors not only coexisted with chimpanzees but may also have interbred with them for hundreds of thousands of years until we finally diverged.[13] Learning precisely how, at the genetic level, we are similar to or different from our nearest animal relative might make it possible to accentuate such distinctive human characteristics as symbolic thinking and our ability to reason morally.[14] In the distant future, our understanding of just how we finally emerged from earlier primate species could lead to the emergence of a new, transhuman species, one as far beyond us as we are beyond chimpanzees.

Let's assume that in the foreseeable future we will vastly develop our ability to understand the meaning of genomic sequences and be able to modify them at will at the cellular level. How could we use these powers to produce genetically modified human beings? In replying to this question, we should recognize first that, to a limited extent, we are already producing human beings to order. Since the mid-

1990s tremendous progress has been made in refining and clinically applying the gene selection technology known as preimplantation genetic diagnosis. This combines in vitro fertilization (IVF) with molecular genetic analysis to permit parents to select early-stage embryos free of a known familial genetic disease. At present, researchers have found more than 1,250 disease-related gene mutations that can be identified and potentially avoided by means of PGD.[15]

Mark Hughes is one of the pioneers of reprogenetic and PGD research. Following a distinguished academic medical career at Baylor University, the National Human Genome Research Institute of the NIH, and Wayne State University, Hughes went into business for himself and now heads the Genesis Genetics Institute in Detroit, with one of the world's leading PGD programs. I served with Mark Hughes on the NIH's Human Embryo Research Panel back in the early 1990s and grew to respect his pioneering work in helping couples at risk for transmitting serious genetic diseases have healthy children. In moral terms, this seemed to me a great advance over the best existing alternative, amniocentesis. Amniocentesis involves the extraction of fetal cells from a pregnant woman's amniotic fluid at fifteen to sixteen weeks' gestation. If the fetus tests positive for the disease, the woman (or couple) then faces the difficult moral decision to terminate the pregnancy at a fairly advanced stage. PGD, in contrast, takes place several days after conception and before the embryo is ever transferred back to a woman's womb.

Hughes's research was just starting when I first knew him, so I decided to pay a visit to his laboratory in Detroit to see how far the technology has progressed in the past decade. The Genesis Genetics Institute's laboratories occupy a suite of offices in an urban medical center. As Hughes showed me around the facility, I learned that each day, Federal Express couriers deliver tiny vials filled with one or two cells that have been extracted from early-stage embryos in infertility laboratories around the country. The embryos are produced by means of IVF for couples who know they are carriers of one or another severe genetic disease and who may already have had (and lost) a child suffering from the condition. This includes people with cystic fibrosis in their families, or Fanconi anemia, a fatal blood disorder, or, worst of all, Lesch-Nyhan Syndrome, one of the most terrifying diseases imaginable. Children born with it cannot process uric acid properly, so it builds up in their tissues. In the first year of life this leads to symptoms like severe gout, poor muscle control, and moderate retardation. Because of their metabolic problems, these youngsters compulsively bite their lips and chew their fingers. Eventually they have to be placed under restraint. Even so, they will gnaw their lips until, often as a result of infection and the kidney damage done by the disease, they die at a young age.

During the early 1990s, when Hughes was trying to refine the PGD technique for routine clinical use, the challenges were extraordinary. Researchers had already become skilled at using micromanipulators to extract a single cell (or "blastomere") from an eight-cell embryo, a technique known as "single cell blastomere biopsy." Since the embryo would soon die unless transferred to a womb, this left the researchers with at most a day or two to perform the genetic tests, and there was never much DNA to work with.

During my visit to the institute in Detroit, I was impressed with how much progress Hughes has made using PGD. His institute now performs hundreds of tests a year. Because of the expanding knowledge of the genetic basis of disease, he can test for dozens of disorders. As we walked through the facility, Hughes showed me a room with shelves holding banks of thermyl cyclers, each about the size of a desktop printer. These devices use a technique known as polymerase chain reaction (PCR) to amplify the tiny samples of DNA received by the laboratory, making DNA sequencing much easier than before. In another area, two automated sequencing machines, each no bigger than a small photocopy machine, spew scrolls of paper with the precise nucleotide sequence from the region of interest in each embryo. By examining the color-coded rows of As, Cs, Ts, or Gs, Hughes and his colleagues can determine whether each embryo tested carries a harmful sequence. This is done rapidly enough to determine which of the several highly perishable embryos available to the parents can be used. Within a day, the results are faxed back to the infertility program. Each embryo's number is listed on a sheet. Next to it is a notation indicating whether it is affected or unaffected by the disease. Only unaffected embryos are used to start a pregnancy.

Almost all of the conditions tested for at the institute represent serious genetic disorders that the child-to-be faces. One exception is the immune-system or HLA (human leukocyte antigen) profile of some embryos. These embryos have been deliberately conceived so that the resulting baby can provide a matching bone marrow transplant for an older sibling suffering from a disease like X-SCID or Fanconi anemia. They are often called "savior children" because without them the existing child would die. The institute program routinely tests embryos for HLA status when parents request it. Defenders of testing believe that the goal of saving lives, even though it is the life of a sibling rather than the tested individual, makes this a legitimate medical use of PGD. Critics fear that it is a first step down the slippery slope to gene testing and gene enhancement for nonmedical reasons.

Everyone has limits. In Hughes's case the limit is sex selection for nondisease conditions. Over a beer that evening following my visit to his lab, Mark was vehement about his refusal to provide tests for sex. "Sex is not a disease," he said. I was impressed by how much he, a world-class genetic researcher, remains a physician dedicated to preventing or curing disease. As we shall see, not all PGD program managers share his understanding of the appropriate uses of their services.

Regardless of critics' objections, there are inherent limits to PGD as a tool for human gene modification. Even if Mark Hughes wanted to produce a baby with some novel genetic characteristic (which he does not), PGD limits him to the choices available in the embryos produced through IVF. In HLA testing, for example, on average only one in five embryos is likely to match the affected child. If a couple's IVF efforts yield five viable embryos, therefore, it is possible that none of them will have the appropriate constellation of HLA genes needed to help the sick sibling. The IVF-PGD process currently permits only selection, not modification. But research is now under way to take reprogenetic medicine beyond PGD to produce genetically modified human beings.

One approach involves a human application of the technology used in connection with transgenic mice. Beginning with a population of stem cells culled from a human embryo, researchers could use homologous recombination to alter the cells' gene sequence in a desired way. Cells that took up the change would then be injected back into that embryo or another one produced by the same parents.[16] Like the mice, the resulting child would be chimeric: some of its cells would have the good gene and some the bad. In many cases this would prevent the disease, because in recessive genetic disorders like X-SCID and cystic fibrosis, the cause is a nonfunctioning gene and a missing protein product. A chimeric child with at least some cells producing the missing protein might be very healthy. If you think it odd that a child could have two genetically different cell populations in its body— in effect two genomes—note that a low incidence of chimerism occurs naturally when two different early-stage embryos fuse in the womb. A small percentage of human beings are chimeras. Usually this is not discovered unless a person undergoes genetic tests, when it is found that the individual has cells of one genetic identity in, say, the bloodstream and the cells of another in the sex cells.[17]

In some cases, chimerism will not work. The functioning of the gene may be harmful, so its presence in any cells in the body is unacceptable. This is true for some genetic conditions like the gene mutation that causes Huntington's disease. The genome of people with this condition reveals a long series of nonsense repeats of three DNA letters. For unknown reasons, this interferes with a gene responsible for key brain functions. In the fourth decade of life or later, Huntington's disease sufferers start an inexorable downward course marked by tremors, weight loss, depression, and, finally, total neurological collapse. How can we correct an embryo to produce an individual whose cells are free of this defect? The kind of forced cross-mating of chimeras used to produce transgenic mice cannot be done in human beings—at least for moral reasons—and even if it could, it would not benefit a couple seeking a healthy child.

Here is a role for human cloning (or nuclear transfer) technology. Although people tend to think of cloning in terms of its ability to create genetic copies of a living person, it is also a powerful tool for gene modification.[18] A researcher could begin with a line of stem cells derived from an embryo made from the parents' sperm and eggs. Homologous recombination could be used to repair the gene defect. But instead of injecting several of these cells into an embryo to make a chimera, the scientist could take one of the corrected stem cells and, using the tools of micromanipulation widely available in infertility laboratories, insert it into one of the mother's eggs from which the nucleus had been removed. This "reconstructed embryo" now possesses a full complement of forty-six chromosomes. Given a mild electric shock, it would begin to divide and grow just like an egg that had been fertilized by a sperm. It could then be transferred back to the mother's womb for development until birth. This is how Dolly the sheep and many other cloned mammals have been produced. In this case, the result would be a human infant that has the corrected gene sequence in every cell of its body.

Mammalian cloning is currently far too unreliable and risky a process to be used to accomplish gene modifications. It took almost 300 eggs and scores of cloned embryos to produce Dolly in 1997, and nearly a decade of trying has not produced much better numbers. Many cloned animals die during gestation or soon after birth as a result of subtle genetic errors, and cloned individuals may also suffer serious health problems. But, like all the technologies discussed here, cloning is moving forward. Its progress may converge with the other capabilities we are examining to forge an entire system for gene modification. In 2006, Ian Wilmut, the scientist who led the team that cloned Dolly and who opposes reproductive cloning, argued for the use of this technology to avoid the birth of children with genetic defects.[19]

A final approach to putting modified cells into babies is worth mentioning, even though it raises special problems of its own, because it is potentially very safe: the direct genetic modification of sperm or eggs. Just a few years ago, a team of Japanese researchers succeeded in injecting small DNA sequences directly into the testes of mice. When these males were mated with normal females, the result in 50 percent of the cases was transgenic pups.[20] The great advantage of this approach is that a male produces millions of sperm, making it possible to use one or another gene modification technique and then examine the large population of sperm cells for those in which the modified DNA has been taken up. Sperm modification provides a way of preventing the transmission of harmful genes in families where the father carries a disease-causing mutation. To avoid disease carried by the mother, or to be certain that there is enough of a modified gene to confer a new, beneficial trait on the resulting child, the parents might also choose to modify the mother's eggs. The limited number of oocytes that a woman can provide, even when she is superovulated with powerful reproductive medications, complicates use of this technique. As we shall see, major developments are under way that promise to remove that limit.

One of the great problems facing all the approaches to gene modification mentioned so far, with the possible exception of sperm manipulation, is that they usually start with IVF. The careful insertion of new DNA sequences and the verification of their uptake usually require working with embryos, and that means gathering numerous eggs from a woman and arranging for in vitro fertilization under laboratory conditions. But IVF is a costly procedure usually resorted to only by infertile people who are willing to pay the eight to ten thousand dollars that clinics charge for a single cycle of drug stimulation, egg retrieval, incubation, and transfer of the embryos into the womb. Few people are able or willing to undertake this costly and uncomfortable procedure for gene modifications that produce only modest benefits for their offspring.

Within one or two decades, however, two new technologies could make IVF the way that many babies are conceived. The first technology is egg freezing; the second is in vitro oocyte maturation (IVM). Taken together, both technologies may change the way people start their families.

We still cannot easily freeze human eggs. We hear about frozen sperm and frozen embryos, and we assume that female reproductive cells can also be frozen. But eggs are among the largest cells in the human body—they can be seen with the naked eye as tiny dots the size of the printed period at the end of this sentence. The eggs are filled with water. When they are frozen, the water forms crystals that disrupt the egg's delicate structures. Sperm and embryos also contain water, but these cells are much smaller (embryonic cells reduce in size as they multiply within the perimeter of the fertilized egg). The reduced amount of water is more easily suffused with the antifreeze (cryopreservant) fluids used in the freezing process.

Within the past few years, researchers in Italy and Japan have developed new techniques for egg freezing. Apparently, one of the keys is sugar. By adding just the right concentration of sucrose to the cryopreservant and by alternating several cycles of fast and slow cooling, Italian researchers have been able to produce eggs that, when thawed, are able to be fertilized at almost the same rate as fresh eggs and that have been used to start healthy, successful pregnancies in human beings. These techniques have now crossed the Atlantic and are being offered on an experimental basis in some of the leading infertility programs in the United States.[21]

Egg freezing is a boon to women suffering from cancer who have to undergo chemotherapy that may damage their ovaries and reproductive ability. It permits them to bank a supply of healthy eggs with which to have children after they recover their health. Many also see egg freezing as a great breakthrough for the millions of women whose career and reproductive decisions now take place to the noisy ticking of the biological clock. It is not uncommon today for young women seeking careers in medicine, law, and higher education to find themselves in their mid-thirties before they have the opportunity and freedom to conceive their first child. Embryo freezing cannot help these women: when they are young, they often have not married or found a partner. By the mid-thirties, however, normal aging causes a woman's eggs to undergo a steep decline in quality. This degradation can cause reduced egg viability, as well as birth defects like Down syndrome. Many infertility programs today work with women who, having chosen to build careers or to wait until Mr. Right comes along, find that they experience problems when they try to conceive. While feminist critics rightly blame some of these problems on male-defined cultural patterns, including educational and business institutions that expect women to put in long years of apprenticeship before earning career success, it is unlikely that these institutions will change soon.

Egg freezing is a partial solution to this problem. Once perfected, it could permit a young woman to put aside a store of eggs to use when she is in her thirties or forties. It is the age of the eggs—not the mother—that impairs their viability. Since freezing halts the aging, and since most women can safely bear a child into their forties, this is a very attractive way of stopping the biological clock. However, there is one serious obstacle: fertilization requires a ripe egg, but most women produce only one or two in each monthly cycle. To increase the supply of ripe eggs, infertility doctors must superovulate the woman by administering potent drugs, which cause most of the discomfort and costs of the egg retrieval procedure. How many women in their teens or early twenties are going to spend thousands of dollars and undergo weeks of drug stimulation to produce, at most, a dozen eggs? Since some of these eggs will not be fertilizable by the time they are thawed years later, the procedure represents a costly gamble.

Enter in vitro egg maturation (IVM). This technology, already being attempted with mouse eggs, mimics the process that takes place each month inside the ovaries when, from the stock of hundreds of thousands of immature eggs, the body chooses one or two to ripen for fertilization. When clinicians can mimic this process in vitro, the world of reproductive medicine will change overnight. Then, without the expense or difficulties of drug stimulation, a young woman could undergo a onetime, outpatient biopsy and put aside a small slice of ovarian tissue containing hundreds or thousands of tiny, immature eggs for freezing. When she is ready to start her family, a few of these eggs could be thawed and matured in vitro to the point where they could be fertilized with her partner's sperm. If the procedure does not work, she would still have an ample supply of eggs on hand to try again.

Once egg freezing and IVM are available, many women will find these techniques attractive. It might become a rite of passage for mothers to take their teenage daughters to the doctor to put aside a store of eggs. As some have quipped, these technologies could lead to a world where sex is for fun and reproduction usually takes place in the laboratory. If even a minority of women avail themselves of these new opportunities, however, the door is opened on a brave new world of genetic modification and gene enhancement. In this world, many pregnancies will begin in a laboratory where clinicians have many eggs, sperm, and embryos for each couple. As selection via PDG becomes much easier and more routine, it opens the way to the deliberate modification of genetic material using some of the techniques explored here. Gene identification and targeting combined with egg freezing and IVM—all technologies not far from deployment—move us directly into the world of gene enhancement.

CHAPTER 3

Drawing Lines

Reginald Crundall Punnett, a British geneticist who worked during the first half of the twentieth century, spent much of his life studying sweet peas and domestic fowl. His name lives on not so much for that work but for a visual aid he invented that geneticists still use to communicate parents' chances of passing on a classic Mendelian trait (or disease) to their children. A Punnett Square consists of a large rectangle subdivided into four equal compartments, two above, two below. A mother carries two versions of a genetic trait, one on each of her two chromosomes. These versions are inscribed above the top two compartments, with the dominant trait—the one likely to manifest itself in the offspring if even a single copy is transmitted—to the left. The father's two versions are inscribed next to the two left compartments, with the dominant one above the other. Filling each box in the square with the corresponding maternal and paternal versions of a trait makes it easy to see the possible combinations of dominant or recessive gene variants that could crop up in any four of the parents' offspring.

Figure 1 is a classic Punnett Square for a genetic trait inherited in Mendelian fashion, in this case eye color. Here the mother and father are both hybrid for brown eyes; that is, each has a gene for brown and blue on the two chromosomes, but their eyes are brown because the

	Mother's Dominant Trait = Brown	Mother's Recessive Trait = blue
Father's Dominant Trait = Brown	Brown + Brown = Brown	Brown + blue = hybrid Brown
Father's Recessive Trait = blue	Brown + blue = hybrid Brown	blue + blue = blue

Fig. 1. Inheritance of Dominant and Recessive Traits (Eye Color)

trait that is dominant prevails in a hybrid mix. "Brown" is capitalized in the Punnett Square to show its dominance. The square reveals that out of every four children these parents have, the odds are that one child will have brown eyes, with two copies of that gene, one from each parent; one child will have blue eyes; and two children will be brown-eyed hybrids like their parents.

Do geneticists naturally think in Punnett Squares? I doubt it. It is purely accidental but interesting that the major ethical choices presented by human gene modification can also be put in a Punnett Square.[1] Figure 2 is such a subdivided square. Inscribed next to the two left compartments are the two choices that relate to the biological target of an intended gene intervention, with the shorter-lived alternative on the top. Is the aim to modify cells in the body of a living individual (somatic cells)? These changes will last as long as the cells are alive but will vanish when the cells (or person) die. This type of intervention is called "somatic cell gene therapy." Or is the aim to change sex cells

	Treatment	Enhancement
Somatic Modification	Somatic treatment	Somatic enhancement
Germline Modification	Germline treatment	Germline enhancement

Fig. 2. Choices in Gene Modification

in order to modify a genetic trait that could be passed on to descendants? This can involve gene changes before fertilization—changes aimed directly at a parent's sperm or eggs—or after fertilization, modifications in the cells of an early-stage embryo. Such changes are likely to affect all the cells in the resulting individual, including the sex cells. This means that the changes stand a good chance of being passed on to the next generation—and maybe well beyond, to the individual's distant descendants. Modifying the sex cells in this way is known as "germline gene therapy."

Inscribed above the top two compartments are two choices that pertain to the purpose of the intervention. The choice on the left represents treatments aimed at eliminating disease. People who receive these interventions carry disease-causing gene sequences, and the modifications aim at curing or avoiding illness. On the top right side we have a choice of interventions that involve people who are not sick at all. Modifications here aim at improving or enhancing normal functioning. The goal is to make people "better than well."

We can identify some of the moral options before us if we fill each of the four compartments of the square. Like the windowpane it resembles, this square affords us a glimpse into our possible genetic future. Some see a bright future in every quadrant. Others are troubled, even horrified, by what they see in one or two of the panes.

Almost no one objects to the gene therapy efforts represented in the top left quadrant: somatic cell gene modifications aimed at curing or treating disease in an existing person. While this may be risky and dangerous, if the disease is serious enough, as in the case of the X-SCID research, it is usually worth trying.[2] In the United States, a federal body, the NIH's Recombinant DNA Advisory Committee, regulates somatic cell gene therapy research. Since the early 1990s, the RAC has approved many somatic cell gene therapy trials. Successes and failures in this quadrant are part of the familiar and relatively uncontroversial story of medical research aimed at curing disease (although some people have objected to the use of the term " gene therapy" for this research, preferring "gene transfer" because so far there have been few therapeutic benefits from it).

The lower left quadrant, germline gene therapy, is far more controversial. It is still prohibited by the RAC. In 1998, W. French Anderson, a leading researcher who made some of the first attempts at somatic cell gene therapy in the early 1990s, proposed an experiment in which he would insert DNA sequences into fetuses suffering from serious inherited blood disorders. Anderson believed that the vectors carrying the DNA were more likely to be taken up by the smaller fetal cells. He also knew that the diseases would probably kill some of the fetuses before they were born unless something was done. Anderson was not deliberately trying to change the fetus's sex cells, but after some deliberation, the RAC concluded that this could happen. It judged the experiment to be germline gene therapy, called for further studies, and indefinitely postponed approval.[3] In the future, it might be possible to perform therapeutic fetal "gene surgery" in the womb without implicating the sex cells. Such therapies would likely meet broad approval.

Fears about germline gene therapy are easy to understand. If a clinician makes a mistake in somatic cell therapy, as in the case of the Necker Hospital X-SCID experiment, the person being treated may die. But if a clinician makes a mistake in germline gene therapy, the clinician has created a new genetic disease that could be passed on from generation to generation, affecting uncounted numbers of people.[4] It is as though the clinician inadvertently introduced a new form of cystic fibrosis or sickle cell anemia into the human population. It is true, of course, that curing or treating someone for a genetic disease by means of somatic cell gene therapy can also have long-term genetic consequences for society. By allowing individuals who would have died at a young age to survive and reproduce, somatic cell gene therapy may perpetuate a disease condition or increase its incidence.

The upper right-hand quadrant, somatic cell gene enhancement, represents the use of gene transfer technology to provide some added benefit to an individual who is otherwise perfectly normal. The best example of this is gene doping in sports, although this category could fill up quickly with many other types of biomedical "improvements." Because of the relative balance of risk to benefit, this quadrant raises difficult ethical questions. All gene therapies carry some degree of risk, whether from misinsertion of sequences or from immune reactions to the DNA vectors that are used. Is it ever morally right for medical professionals to subject someone to serious risks when the aim is not to remedy a disease but merely to provide an added benefit? Doesn't this violate a key tenet of medical ethics: "Above all, do no harm"? Some people believe so, but in a world where cosmetic plastic surgery is widely practiced and widely sought, the answer is far from clear.[5]

	Treatment	Enhancement
Somatic Modification	X-SCID gene therapy	Gene doping in sports
Germline Modification	Removing X-SCID (or CF or sickle cell anemia) from a family line	Conceiving a "superathlete"

Fig. 3. Typical Genetic Interventions

The fourth and final quadrant involves what are called inherited germline genetic enhancements. These raise the same ethical questions as somatic cell enhancements do, but all the further risks to our descendants associated with germline interventions are added. The high degree of risk has led regulatory bodies like the RAC to conclude that research of this sort should be banned for now. Thus the lines of our Punnett Square are also fences. Many people think that, at least for the foreseeable future, research and practice should stay inside the upper left quadrant. Figure 3 depicts typical genetic interventions that could reside in each of the compartments of our treatment-enhancement–somatic-germline Punnett Square.

Developments in medicine and society are pressing us to jump the fences and allow gene modifications in all the compartments in the Punnett Square. One source of the pressure to expand gene interventions is the blurring distinction between treatment and enhancement. On the surface, the two seem very different. Treatment aims at bringing people up to a level of relative normalcy, enhancement at making them supernormal. In July 2003 a U.S. Food and Drug Administration panel ruled that adolescents whose height is two and a quarter standard deviations below normal for their age are eligible for treatment with human growth hormone (hGH). The ruling was controversial because it did not require a clear diagnosis of hormonal deficiency, as had been true before, just a measurement of height.[6] A growing literature also indicates that not all very short boys (or very tall girls) are significantly impaired in psychological terms.[7] Nevertheless, the ruling made sense in terms of medical care. Extreme shortness does increase a child's vulnerability to social and psychological problems. In addition, diseases can often be treated without understanding what causes them; for example, although no one fully understands why some people fall victim to schizophrenia, we treat the disease with available drugs because of its abnormality and the suffering it causes.[8] The FDA panel applied this logic to extreme shortness when it ruled in favor of hGH therapy. But although we are willing to regard hGH as a medical treatment for a child because it eliminates the physical inconveniences of extreme shortness and the teasing and bullying often accompanying it, we are far less comfortable making the leap to enhancement by prescribing hGH to a boy of normal stature whose parents want him to become a star basketball player.

This distinction between treatment and enhancement is enshrined in many of our social practices governing medicine. We honor doctors who provide restorative treatments, and we believe that insurance companies should cover the costs of the care. But we are less enthusiastic about doctors who offer enhancements, and we permit insurers to refuse coverage for such procedures. A woman who loses a breast to cancer has no problem receiving insurance coverage for reconstructive surgery. But breast implants for healthy women who want to improve their appearance are never covered. Doctors who offer them and the patients who seek them may be subjected to ridicule.

When the procedures prove risky, as some maintained in the case of silicone implants, we are quick to prohibit them.[9]

As rigid as these fences seem to be, they fall down for two reasons: one pragmatic and one conceptual. The pragmatic problem arises whenever a drug or treatment becomes available for legitimate medical purposes. At that point, it is very hard to stop the expansion of its use for enhancement. Partly this is because, under U.S. law, once a drug has been approved for a class of conditions, a doctor can also prescribe it for other purposes. This is known as "off label" use. For example, Botox (botulinum toxin) was initially introduced as a muscle relaxant for serious medical conditions. Now it is widely used by plastic surgeons and others as an all-purpose wrinkle remover. Diversion is another reason for the expanding use of a treatment. Once a drug (or procedure) becomes available, it is hard to prevent its getting into the wrong hands. Viagra was initially introduced to help men who suffered from erectile dysfunction following surgery for prostate cancer. It is now widely used on college campuses and in the gay subculture by people hoping to become sexual athletes.

The conceptual problem arises the moment we consider that some of our most valuable medical interventions *are* enhancements. Vaccines are a leading example. Almost no one is naturally immune to smallpox, polio, measles, whooping cough, or any of the other diseases that we vaccinate against. When we are inoculated, the DNA in our white blood cells undergoes irreversible genetic changes, initiating the synthesis of antibodies to many viruses and bacteria. Vaccinations make us superhumans, but no one ridicules enhancements of this sort. In most places in the United States and other industrialized countries, a child cannot enter school unless he or she is vaccinated.

This tells us that the sharp distinction that our Punnett Square makes between treatment and enhancement is misleading. As figure 4 shows, there is an intermediate zone between the two: prevention.

	Treatment		Enhancement
Somatic Modification	Somatic treatment	P R E V E N T I O N	Somatic enhancement
Germline Modification	Germline treatment		Germline enhancement

Fig. 4. The Place of Prevention in Genetic Intervention

Like treatment to its left, prevention belongs to the realm of disease and illness, those subnormal bodily states that either cause or signifi-cantly increase our risks of suffering pain, disability, or death.[10] The dangers that prevention aims at lie in the future. The goal is to surpass normal levels of functioning now to prevent them from ever occurring. In a way, preventions are a kind of enhancement aimed at *maintaining* normalcy. In this respect, they are deeply related to the traditional goals of medicine. They differ from what I would call "pure enhance-ments," which have nothing to do with forestalling disease or disabil-ity. Pure enhancements aim at gratifying the wishes of normal and healthy people for improved performance or superior capabilities.

Understanding how important prevention is in our medical and moral thinking lowers the barriers to gene interventions erected by our Punnett Square. Take the category of somatic cell gene enhancements (our upper-right quadrant). So long as we confine our thinking about this to pure enhancements, like gene doping in sports, that have noth-ing to do with preventing disease (and even risk causing disease), it is easy to see why we are uncomfortable with them. But what if a somatic gene enhancement protects against disease? Not long ago, scientists researching HIV/AIDS made an important discovery. Working with prostitutes in Kenya, they found that a small number of the women had not contracted AIDS even though they had been exposed repeatedly to the virus that causes the disease. Eventually, this discovery led the scientists to identify a variant form of a gene for a receptor on the surface of human cells. (Each of our genes has two copies, and geneticists use the term "allele" for a variant version of one of them.) The gene involved is known as CCR5, and the receptor it builds is the docking point for the virus. HIV cannot infect people who lack this receptor, like the healthy Kenyan prostitutes, even when a large dose of the virus directly enters their bloodstream.

Imagine now that we could invent a gene vaccine that artificially mimics what this allele does. We could inject people with vectors into which were stitched DNA sequences that disrupt the functioning of the normal CCR5 gene. The DNA sequences would render the cells insusceptible to AIDS infection. In view of the devastation being wrought by HIV/AIDS around the world, with millions of people dying annually and tens of millions already infected with the virus, would anyone object to the injections? With just one injection, an HIV/AIDS gene vaccine could replace the costly and often ineffective antiretroviral therapies now in use. Would anyone say, "We should not introduce this vaccine because it crosses the line between treatment and enhancement and represents a step into somatic cell gene enhancement"?

Let's go further. Since cells in the body die and are replaced by new, unmodified cells, even a somatic cell HIV/AIDS vaccine might require repeated administration. Imagine instead that we could invent a gene therapy suppository that a woman could use before intercourse, with or without her partner's consent. This would infect her sexual partner's sperm with a harmless virus carrying sequences to disable the CCR5 gene. As a result, any child the woman conceives would have that change in all the cells of its body, including the sex cells. The result would be a generation of children who were naturally more resistant to HIV infection and who would likely pass that resistance on to their children. Again, would anyone object? I doubt it. It is hard to imagine someone saying, "We shouldn't permit this because it represents germline genetic transfer for enhancement purposes." So, once we are involved in efforts to prevent or mitigate disease, none of these barriers is impermeable. We can anticipate widespread support for gene modifications that prevent the future occurrence of disease or disability. This could include somatic cell enhancements like the gene vaccines I have mentioned, or even germline modifications that promise to eradicate a harmful gene or gene variant once and for all.

Another example of a desirable germline therapy might be an intervention aimed at preventing sickle cell disease, a serious blood disorder caused by a single-letter genetic misspelling in the gene that produces the protein for hemoglobin, the oxygen-carrying component of red blood cells. (The DNA chemical letter A, adenine, is changed to T, thymine, altering one of the amino acids that make up the resulting protein.) Normal hemoglobin causes the red blood cells to assume an oval shape, but the defective hemoglobin gives them a sickle-like appearance. These irregularly shaped cells jam and stack up in the narrowest blood vessels of the circulatory system, depriving organs and tissues of blood. This causes periodic episodes of pain, and damages tissues and vital organs, including the retinal cells of the eyes. In severe cases, the disease leads to premature death.

Sickle cell is a recessive disorder. Only someone who entirely lacks the normal protein product because both copies of the hemoglobin gene are defective is likely to suffer from the disease. It usually results when two carriers of the disease, each with one normal and one defective allele, have a child. In such cases, a Punnett Square tells us that 25 percent of their offspring (on average, one in every four of their children) will have two copies of the defective gene and suffer from the disease. Imagine now that we could inject the father's testes with DNA vectors that could correct the misspelling, changing T to A. He would cease being a carrier. Given the genetics of this Mendelian disorder, no matter what the genetic status of the mother, none of their children would ever be at risk for the disease.

We could go further. If the mother was affected by the disease or was a carrier, there would be a chance of transmitting the disease to future members of the family even if the father's genes were corrected. But there is a way of totally eliminating the gene defect from a family line. Recently, a team of scientists at the University of California in San Francisco used homologous recombination to repair a population of mouse stem cells that had had a defective human sickle-cell gene inserted into them.[11] Theoretically, the nucleus of a repaired human stem cell like this could be used in a cloning procedure to produce a child that neither suffered from nor carried the defect. Although cloning is still too risky a procedure to be implemented for gene therapy, it is one of the ways we could someday entirely eliminate sickle cell disease, cystic fibrosis, and other serious disorders that continue to appear in some families—using the very germline therapy that now evokes apprehension and resistance.

Some have argued that germline therapy is never a good idea. Noting the long-term risks, they point to a series of alternatives. One is improved drug therapy for known diseases. For example, in the past few years, hydroxyurea, a drug approved by the FDA to treat certain types of leukemia and other cancers, has been found to reduce the severity of sickle cell disease in many patients. There are other dramatic new treatments likely to be discovered. Somatic cell gene therapies may also help. In the case of sickle cell disease, genetically modified stem cells could be injected into a patient's body to moderate the course of the disease. But all these approaches have problems. They require repeated applications; they may not repair all the harm done by the bad gene, especially harm done during fetal development; and, as in the cases of Jesse Gelsinger, killed by an immune reaction to a viral vector, and the X-SCID children, doomed because bone marrow therapy failed, none of these somatic cell approaches is entirely free of risks. Why keep repeating dangerous halfway therapies in each generation if you can get rid of the disease once and for all through germline gene therapy?

One alternative avoids these problems: preimplantation genetic diagnosis. After going through an IVF process followed by PGD, a couple could ask their clinicians to set aside all the embryos having one or more copies of the defective gene. As in the case of gene repair, the inherited defect would be eliminated from the family line. With this alternative already at hand, some ask why we should ever undertake gene repair approaches with the risk of making mistakes that could be passed on to uncounted future generations.[12] If we can select, why become involved in the potentially dangerous business of germline gene modification?

One answer to this question is that in some cases a couple may not produce enough embryos to permit selection. There are also certain very rare instances where both parents suffer from a recessive disorder, and all of their embryos carry two copies of the defect.[13] This is beginning to happen today as aggressive new therapies permit more young people with cystic fibrosis to survive and reach reproductive age. If two of these individuals meet, marry, and want to have a child, none of their sex cells will carry a gene unaffected by the disorder. When somatic cell gene therapy succeeds, this problem will only grow with time. IVF and PGD are also technically demanding and costly procedures. Although gene modification followed by cloning may be even more so, one form of germline gene therapy, the direct manipulation of the father's sperm, may prove to be an easy and efficient way to reduce the risks of genetic disease transmission.

With some genetic conditions, too, the problem is not a bad DNA sequence but a missing one. In these cases, selection will not work, since there will be no embryos with the desired sequence. The parental genes have to be actively modified to produce the desired result. This is true for some gene disorders, but it is also usually the case where gene enhancements are concerned. Here the purpose is not to mute or eliminate a malfunctioning gene but to actively introduce into the genome DNA sequences that provide new or heightened functioning.

Finally, I should note that for religious reasons many people oppose PGD because it involves the creation, selection, and discarding of early human embryos. Roman Catholic teaching holds that the embryo should be regarded as fully human from conception onward, and it prohibits creating human embryos in vitro or discarding them after they are created, a view shared by many evangelical Protestants. (Orthodox Judaism, although it opposes abortion except when needed to save the mother's life, accepts both IVF and the discarding of early embryos.) Catholics and Protestants who oppose PGD are likely to regard the direct modification of sex cells or gene therapy on early-stage embryos or fetuses as morally preferable to selection and discard as a way of addressing genetic problems. In fact, as we shall see when we turn to religion in chapter 7, Pope John Paul II has stated that *therapeutic* gene surgery is compatible with Catholic moral teaching so long as it does not involve prohibited methods like masturbation or IVF that separate sex from procreation.[14] Gene modifications, including germline modifications that aim at sperm or eggs in the parents' bodies or that are targeted at the early fetus in the womb, could pass this test. One of the ironies of reproductive medicine today is that conservative religious teachings may be a driving force for the introduction of a revolutionary new germline genetic approach to preventing illness.[15]

The natural association between gene modification and enhancement leads some to argue that germline gene therapy makes sense only for pure enhancement purposes, where individuals want to introduce new and "improved" gene sequences not related to remedying disease and not present in the parents' genomes. I do not believe this is entirely true. There are ample reasons why germline gene therapies might make sense in future efforts to eradicate genetic diseases. Our intermediate category of disease prevention shows where selection will not work and new sequences are needed to change the genome in ways that make people more resistant to illness. An HIV gene vaccine is an example. These directions suggest a place for disease-related somatic and germline interventions in our future.

This still leaves the fourth quadrant of our Punnett Square, where pure enhancements reside. The goal here is functioning above and beyond the normal in ways that have nothing to do with curing or preventing disease. Can this ever be justified? Is it wise or morally permissible to meddle with the human genome just to offer some healthy people advantages that they would otherwise lack?

Parents would deny what Canadian researchers discovered in 2005: attractive children receive better care than ugly ones. A research team followed more than four hundred parents and their two- to five-year-old children around supermarkets. The researchers had previously judged the children's attractiveness on a ten-point scale and come up with an unsettling finding: homely children were more neglected and were allowed to engage in more dangerous behavior.[16] Only 1.2 percent of the homely children were buckled into the shopping cart, compared with 13.3 percent of the cutest ones. When a man was in charge of shopping, none of the unattractive children was strapped into the carts, whereas 12.5 percent of the attractive children were. Uglier children were also allowed to wander more than ten feet away and engage in such risky behaviors as standing up in the cart.

This finding conveys a bitter truth. Beauty may be only skin deep, but people's success in life is deeply affected by how they look.

In their book *Mirror, Mirror: The Importance of Looks in Everyday Life,* Elaine Hatfield and Susan Sprecher marshal a great deal of evidence in support of this conclusion.[17] In one study, more than 400 fifth-grade teachers were asked to evaluate children's prospects for academic performance and career success. Appended to each student's academic record was a photograph of a plain or attractive boy or girl. The photographs and the records were bogus, but the researchers found that, regardless of the record, teachers rated the attractive children's prospects much more highly than the plain ones'. In another experiment, researchers left application forms to graduate programs in psychology in phone booths at the Detroit airport. Clipped to each form were a stamped, addressed envelope and a note to the applicant's father asking him not to forget to mail the application at the airport. The forms contained standardized information: name, address, undergraduate record, and grade point average. Attached to each was a photograph of a homely or attractive college-age man or woman. Would a harried traveler take the time to seal the letter, find a mailbox, and post it? On average, the forms with photographs of attractive applicants were mailed at a one-third higher rate.[18]

Other studies have shown that juries favor attractive defendants over ugly ones. In trials they tend to dole out more severe penalties to unattractive people.[19] This supports the finding that people moralize physical appearance. They seem to believe that "what is beautiful is good, what is ugly is bad."[20] Art reflects this belief by characteristically portraying evil people as ugly or physically deformed.[21]

The importance of appearance has broad consequences for our paths through life. How attractive we are influences others' opinions of us and opens or closes doors to opportunity. Remarkably, it also influences how we regard ourselves. This in turn affects our attitudes and our motivations. Good or bad looks thus become a kind of self-fulfilling prophecy. Hatfield and Sprecher call this the "Pygmalion effect."[22]

What most people seem to desire for themselves and for their children is not so much extreme good looks—striking beauty or handsomeness—as the avoidance of unattractiveness. When a large number of people were shown a sample of photographs that had been prejudged on a scale ranging from the handsome and beautiful to the average, the ugly, and the terribly disfigured and polled on whom they would most like to look like, there was a marked drop-off in preference only at the ugly end of the scale. Most people were content with "average" looks. Other research suggests that to some extent averageness is itself a valued measure of appearance. In a 1990 study, a computerized technique of mathematically averaging multiple faces was used to create composites of male and female adult faces. When subjects were asked to evaluate the faces in terms of attractiveness, the study showed that the more faces that made up the composite, the more attractive was the ranking of the face.[23]

In a society where a pleasing physical appearance is one of the most valued human attributes, we can easily imagine physical attractiveness, or at least the avoidance of markedly unattractive appearance, as high on the list of pure enhancements requested by parents in an age of prenatal genetic interventions. Indeed, attractiveness is the foremost quality sought by people using sperm and egg donors to have children. Although the matter has not been systematically studied, there is anecdotal evidence that parents, regardless of educational, economic, or career background, usually place good looks near the top of their list of qualities when choosing from among donor profiles.[24]

Today it is beyond the ability of genomic science to produce a face to order, although some of the genes that control craniofacial development have been identified. Because genes governing facial development also play a significant role elsewhere in skeletal development, it would be foolhardy at this time to attempt to alter them in an embryo or fetus. The misguided effort to produce a pleasing face might yield a host of skeletal deformities.

But some aspects of our physical appearance will come under genetic control in time. Recently a team of researchers announced that they had found the genes controlling skin color in Europeans and Africans. They were working with one of geneticists' favorite model organisms, the small striped zebrafish, and examining a mutation in some fish that altered melanin levels in its stripes, turning them from black to a pale golden color. They knew that there is a human counterpart (or homologue) for this gene. When they examined human genomic databases for the relevant sequence, they discovered many instances of the same mutation. They further discovered that the mutation was much more prevalent among people of European descent, whereas the sequence associated with the darker-striped zebrafish prevailed among people of African descent. Since it is possible to have one or two copies of either gene, skin tone varies within these two populations, but it appears to be significantly controlled by these two alleles.[25] The research shows how comparative genomics and other tools of molecular biology are greatly facilitating the search for the genetic causes of any human trait. Many of the features relevant to physical attractiveness in our society—facial shape; quality of teeth (there is already solid genetic knowledge of enamel formation);[26] hair, eye, and skin color—will eventually come under our control.

The thought of choosing a child's skin color raises troubling ethical questions. Will dark-skinned African-American parents choose to have lighter-skinned children? And if they do, won't that reinforce Americans' racial prejudices, making life even harder for the darker-skinned children whose parents were either unable or unwilling to change their child's skin tone? The bioethicist Thomas H. Murray puts the question this way: "Does it make sense to deal with racial prejudice by trying to obliterate physical differences? Should we encourage biomedical fixes for complex social problems? Or would we be wiser to deal with the roots of prejudice?"[27] The questions are not new. They were faced in the past by people who chose to undergo rhinoplasty to change a "Jewish nose" and by Asian Americans who wanted to "Westernize" their eyes. In all cases, individuals and families had to ask whether it was better to seek relief from discrimination by fleeing one's differences or by staying and fighting for them. Both strategies have their defenders, and there are good moral arguments on either side.[28] Ultimately, the decision is highly personal, and one for which parents will probably come down on both sides.

If the genetic modification of children before they are born or even conceived ever becomes accepted practice, we can also imagine that parents will be interested in influencing a child's height and weight. Parents' concern with height is already shown by the widespread use of human growth hormone for children who are measurably shorter than normal. Thousands of short children with no identifiable hormonal deficiency are currently being diagnosed with "idiopathic short stature" (ISS), a name meaning, literally, that the cause of the condition is unknown. Treatment requires daily injections of hGH into the child's abdomen over a period of several years at a cost of fifteen to twenty thousand dollars annually, depending on the child's needs.[29] Many children now being treated for ISS probably have genes that substantially contribute to their condition, which suggests that short stature is a natural candidate for germline genetic intervention. Why should parents who know that they carry genes for shortness wait until after their child is born and then expose the child to a prolonged, painful, and costly series of injections when they could correct the problem before birth?

Germline therapy for extreme shortness would probably open the door to parental requests for increased height even when the child was not likely to be abnormally short. This would represent a movement from our intermediate category of prevention to pure enhancement. Research has amply demonstrated that it is a great advantage, especially for men, to be tall. For example, a 2004 study by the economists Andrew Postlewaite and Dan Silverman found that among white American males each additional inch of height translates into a 1.8 percent increase in wages.[30] A survey in 2005 of CEOs of Fortune 500 companies revealed that on average they were six feet tall—three inches taller than the average American male. Fully 30 percent of these CEOs were six feet two inches tall or more. Only 3.9 percent of adult American males are this height.[31] Another recent study concludes that economic discrimination against short adult males is equal in magnitude to racial or gender bias in the workplace.[32] In politics, too, height is a great advantage. In his 1982 book *Too Small, Too Tall* the psychologist John Gillis reported that in the twenty-one presidential elections between 1904 and 1984, the taller candidate won 80 percent of the time. In the whole history of the Republic, he observed, only two presidents—Benjamin Harrison and James Madison—were appreciably shorter than the average height in their day. Most U.S. presidents have been significantly taller than average. Interestingly, although both John F. Kennedy and Richard Nixon were about the same height, when voters indicated a preference for one of the candidates, they were more likely to say their candidate was the taller of the two.[33]

Some parents might want to confer these likely advantages on their offspring, even when there is no reason to believe that the child will be pathologically short. Germline genetic intervention aimed at improving normal height raises questions. Not least is the problem of positional advantage mentioned in connection with sports doping. If everyone tries to have a taller child, will anyone be better off? As we drive up the costs and risks of such interventions, won't there always be some people who are still short? This is a problem, but it need not render useless all efforts to increase height or any other widely desired trait. If many parents attempt to increase their potentially short child's stature, the shape of the height distribution curve will change, with a bulge at the greater height end of the curve and a less bell-shaped distribution overall. The result will be a greater average height and more people near that height—possibly a good outcome for the children involved. In other words, not all quests for enhancement involve the problem of positional advantage. It is only when everyone aims for superiority that the problem occurs. In any event, it is enough to observe that in a world of prenatal gene modifications, improved height is likely to be high on the list of requested gene interventions.

Even more than height, obesity bestrides the line that separates preventive germline therapies from pure enhancements. It is both a health and an appearance problem. The United States is experiencing an epidemic of obesity caused largely by the ready availability of fattening foods and the lack of physical exercise. More than 60 percent of Americans aged twenty years and older are overweight, and one-quarter of all American adults meet the more stringent clinical criteria for obesity. The problem is not confined to the United States. Worldwide, more than one billion adults are overweight, and three hundred million of them are obese.[34] The statistics among children are also disturbing. A Centers for Disease Control "National Health and Nutrition Examination Survey" conducted in 2002 revealed that an estimated 16 percent of adolescents aged sixteen to nineteen are overweight, nearly a fourfold increase since the first such study, conducted between 1963 and 1965.

Being obese greatly heightens a person's risk of suffering from a variety of serious health conditions, including hypertension, high cholesterol, Type 2 diabetes, heart disease, stroke, arthritis, and breast, colon, and endometrial cancer. In terms of appearance, being fat is a painful stigma. Not only do others tend to shun fat people as romantic partners or employees. In addition, they justify these discriminatory attitudes in moral terms. Fat people are seen as responsible for their condition. Research shows that obesity is consistently attributed to laziness and a lack of self-discipline.[35]

In reality, the truth may be just the opposite. Studies of identical twins reared together or apart indicate that much obesity may be caused by hereditary factors.[36] In technical terms, the "heritability" of obesity, the percentage of observed variation among people that is attributable to genes, is very high, somewhere between 50 and 80 percent. Obesity researchers now believe that a person inherits a range of weights that his or her body can comfortably maintain and that dieting and lifestyle changes cannot easily move that range from a midpoint by more than 10 percent.[37] Genetic research is also beginning to uncover some of the specific genes that increase a person's risk of suffering from obesity. In the mid-1990s, Jeffrey Friedman at Rockefeller University in New York isolated one defective gene that made a strain of mice grow immensely obese (the mouse strain is appropriately named Tubby).[38] This gene, which has a human homologue, codes for the hormone leptin, which appears to play a key role in the brain's control of appetite and fat metabolism. Just recently, an international team of researchers using new silicon-chip-based whole genome scanning methods identified a single-letter DNA variant responsible for obesity in 10 percent of people of western European and African ancestry.[39] Summarizing recent genetic research on obesity, one report concludes, "The most important message after nearly a decade of searching for genes that are involved in human obesity is that such genes, perhaps surprisingly, really do exist—despite the key role of modern lifestyles in the current obesity epidemic."[40]

If genes are involved in obesity, the control of weight by genetic means is probably on the horizon. Ideally, we should all eat properly and exercise. But some people wage a lifelong, losing battle against their metabolism: nearly 90 percent of dieters gain back the weight they lose.[41] The rapidly growing popularity in Europe and in the United States of surgical weight-control procedures that tie off portions of the stomach to reduce appetite shows how serious this problem is and what risks people will take to remedy it.[42] Because it is so hard to change the behavioral or environmental conditions that drive obesity, we should not be surprised if some people eventually decide to use prenatal genetics to spare their children this burden in life.

Parents discover Hutchinson-Gilford syndrome only when their two-or three-year-old begins to look odd. The child's hair starts turning gray and falling out. Strange folds and wrinkles appear around the eyes. At ten, arthritis sets in. Girls experience osteoporosis at eleven. The child may experience senility before hitting puberty and die from heart failure or simple old age by nineteen. Like several other diseases of rapid and early aging, known as progerias, Hutchinson-Gilford syndrome has a genetic cause. In 2003 scientists found the specific gene mutation responsible for the disorder. Research is now under way to find a treatment.[43]

The specific genes behind any of the progeria syndromes are unlikely to be major contributors to normal human aging, but they are clues in our growing understanding of the genetic causes of aging. That genes play an important role in longevity has been demonstrated on many fronts. Family studies, for example, show that the male siblings of centenarians are seventeen times more likely than other men born around the same time to live to one hundred. Female siblings are eight times as likely to do so.[44] One theory holds that the reason is that people who reach extreme old age have genetic variations throughout their genome that slow the basic mechanisms of aging and result in a decreased susceptibility to age-associated diseases.[45] This theory is supported by the finding that throughout life the children of centenarians are significantly healthier than the children of people who die at an average age. In one study that compared nearly two hundred children of centenarians with a control group of roughly the same number of children of people who died at a younger age, the centenarians' children experienced the onset of heart disease, high blood pressure, and diabetes five to eight years later than the controls.[46]

A more controversial theory, for which there is also some support, holds that a much smaller number of genes, called longevity-enabling genes, confer protection against the basic mechanisms of aging or age-related illnesses.[47] It has been known since the mid-1930s that animals deprived of food (but given vitamins and minerals) age much more slowly than those fed an adequate diet.[48] One hypothesis is that too much food contributes to the buildup of metabolic waste products, especially free oxygen radicals, which damage cells. Research in mice, flies, and worms has shown that altering the genes on pathways related to food metabolism can radically extend the creatures' life span. For example, one researcher bred a line of fruit flies with a mutation in a gene the scientists called INDY, for "I'm not dead yet." The gene appears to be responsible for forming a membrane that moves nutrients into cells. Disabling the gene created the genetic equivalent of calorie restriction. The mutant flies had average life spans of seventy-one days, rather than the normal thirty-seven days. The long-lived flies also appeared as healthy and fit as normal flies, continuing to reproduce long after most flies without the mutation had died.[49]

Whether aging is caused by a handful of genes or many separate genes working together, we will eventually have a better idea of the genetic factors that contribute to it, and with that knowledge will come power.[50] Some of the knowledge will be used for disease prevention. We can easily imagine gene interventions aimed at preventing or curing tragic progerias like Hutchinson-Gilford syndrome.

Should we go further? If, by genetic modifications to sperm, eggs, or embryos, we could offer our children a life span twenty or thirty years longer than is normal today, should we go ahead?[51] It would be important, of course, to know that those additional years were not likely to be blighted by some of the serious diseases of aging now taken for granted. But, as the fruit fly research suggests, good health in old age may also come under our genetic control.

Age-extension interventions that dramatically increase the number of older people in society raise many ethical questions. A vast increase in the percentage of older people could cause economic dislocations and generate new sources of conflict between the generations. In the next chapter, we shall begin to consider the debit side of the equation. For now, it is enough to recognize that an extended life span may be one of the items on a future menu of prenatal genetic choices.

In 1999, Joe Tsien and his colleagues at Princeton University reported that they had genetically engineered mice with the increased ability to perform learning tasks. The scientists inserted a gene into mouse embryos that augmented the production of a key memory and learning protein. When the mice were run through mazes, the genetically altered ones did much better than normal or "wild-type" counterparts. In other tests, the altered mice proved to be much quicker learners. The researchers named this strain of transgenic mice Doogie, after a precocious character on the U.S. television show *Doogie Howser, M.D.*[52] In the published article on their research, the scientists observed that the gene and protein products they had manipulated are widely conserved across mammalian species, which suggests that the genetic enhancement of such mental and cognitive attributes as intelligence and memory is feasible in all mammals, including human beings. Future research will be needed to see whether this particular genetic finding has applicability to human beings.

Tsien and his colleagues were not trying to lay the genetic groundwork for a race of superintelligent human beings. As they pointed out, many people suffer deficits in some of the neural systems and pathways that they were investigating. Their research opened up new biochemical targets for the treatment of learning and memory disorders. But like most research aimed at identifying and repairing genetic defects, theirs had the consequence of illuminating new ways of enhancing normal performance.

Subsequent tests revealed that Doogie mice not only are smarter than ordinary mice but may also have an increased sensitivity to pain.[53] This finding illustrates the enormous complexity of brain processes and the risks that accompany attempts to improve a child's IQ by genetic means. Furthermore, it is not at all clear that improving memory performance would produce the range of cognitive skills associated with higher IQ in human beings, much less the many aspects of human intelligence that have been identified by cognitive researchers.[54] I shall return to these problems when we look more closely at the limits of genetic interventions. But given the importance of academic performance and test achievement in the highly competitive societies of the developed world, and the amount of time and effort many parents put into promoting such accomplishments, safe, effective genetic enhancements of learning are likely to attract a clientele.

The case for cognitive enhancements becomes stronger when we move back across our line and look at possible preventive interventions. In 2004 and 2005, separate groups of researchers in Sweden and the United States discovered three genes strongly linked to dyslexia. One of the genes, called DCDC2, is active in the reading centers in the human brain. Large deletions in a regulatory region of the gene were found in one of every five dyslexics tested. Another of the genes, called Robo1, is a gene that during fetal development guides connections between the brain's two hemispheres. Defects in this gene do their damage before birth. Postnatal somatic-cell therapies may be less likely to aid in the treatment of dyslexia than would prenatal genetic interventions.

Reading disorders seriously impede career success and contribute to persistent social and economic problems in families where they occur. At a minimum, genetic research in this area can help diagnose problems early to permit more aggressive medical and schooling interventions. Eventually, dyslexia may come to be seen as a genetic disorder that is routinely screened for in prenatal testing and responded to either by embryo selection or by targeted gene modifications.

Interventions aimed at modifying mood or temperament are also in the cards. Consider depression. This is an illness whose expression can range from temporary "bad days" to weeks or months of overwhelming despair that can lead to suicide. A deficit of serotonin in the brain plays a major role in the disorder. Selective serotonin reuptake inhibitors (SSRIs) like Prozac work by slowing the removal of serotonin from regions between the neurons, leaving more of the neurotransmitters in place. Genetic research has shown that several genes affect the availability of serotonin in the brain.[55]

A regulatory region of the gene responsible for serotonin transport appears to have two variants: a short form and a long form. The longer form produces structures yielding more serotonin. In standardized testing, people having the short-form allele exhibit higher scores for depression, as well as heightened anger, anxiety, worry, and pessimism (all of which psychologists subsume under the label "neuroticism").[56] People with the long-form allele are less susceptible to all these negative mood states—as though they are continually on an antidepressant medication. Indeed, the long-form variant has been called "natural Prozac." Remarkably, one of its effects, as with Prozac, is to slightly depress levels of sexual desire. Bearers of the long form may be happier than short-form people, but they may also have less-frequent sex.

No biological system is more complex than the human brain, and the mere thought of tinkering with it makes most people worry. Even this very brief overview of the possibilities makes it clear that well-intended efforts can go very wrong. Like Doogie mice, our learning-enhanced children might be more susceptible to pain. People might experience a brighter mood but pay for it in reduced sexual interest and reproductive success.

The risks are not wholly beyond assessment. We shall see in the next chapter that naturally occurring variants in genes allow us to see in advance what deliberate changes might do. They give a sense of what works in nature and what does not. But even naturally occurring variants are not perfect. The sheer complexity of the human genome limits how confidently we can predict the outcomes of genetic changes.

At the same time, variations in the human genome also tell us that it is not exactly a stable, uniform, and time-proven biological system. It is more like an experiment in progress, or tens of millions of such experiments. The high incidence of genetically influenced diseases and the constant appearance of new gene mutations indicate that not all these experiments are successful. Our pervasive medical and environmental interventions, while saving and prolonging countless lives, have also increased the incidence of harmful gene variants that in previous generations would have been eliminated by remorseless natural selection.[57] The question before us as a species is whether we dare to add novel risks of our own making as we try to improve on nature.

CHAPTER 4

Challenges and Risks

Letitia Blakely was crying. It was hard enough being one of ten NGs (natural genomes) in a high school class of seven hundred PPCs (pre-planned children). Letitia was slightly overweight, her skin was pasty, her hair frizzy, she was bulbous-nosed and weak-chinned, and one of her breasts was larger than the other. Her looks and emotional disarray stood out in a class where just about every other kid had manageable hair, unblemished skin, high intelligence, and a warm and sunny personality. But now, to add insult to injury, Reena Cathcart, beautiful and brilliant Reena Cathcart, was asking her to take the part of a frowsy old lady in the upcoming school play. "You're the only one, Letitia," Reena pleaded. "You know none of the others can. . . . You're simply the only one who can play the old—the older—woman."

Greg Bear's 1989 short story "Sisters" takes place about sixty years in the future, in a world where the genetic enhancement of children has become commonplace.[1] Letitia Blakely's problems are understandable. Because she lacks the subtle, sharp wit of most of her classmates, her dreams of a medical career are futile. She is hopeless at sports. The school actively tries to combat discrimination against its handful of NGs, and her parents have a legal right to sue if she is harassed, but behind her back, peers hurl at her the epithet "TB"—"throwback."

Letitia blames her parents for their decision to not engineer her at conception. In an angry dinner-table conversation, she asks her mother and father, Jane and Donald, why they refused to use the new genetic technology before she was born. "It's not because you're religious," she observes. "No," her mother admits, shaking her head firmly. "Then, why?" Letitia asks. Her father attempts an answer: "Jane and I believe there is a certain plan in nature, a plan we shouldn't interfere with. If we had gone along with most of the others and tried to have PPCs—participated in the boy-girl lotteries and signed up for the prebirth opportunity counseling—why, we would have been interfering."

Letitia is unhappy with her father's explanation. She points out that her mother chose to give birth to her in a hospital. "That's not natural," she says. Donald tries again. "There are limits. We believe those limits begin when people try to interfere with the sex cells."

Letitia remains unconvinced and unhappy, but she grudgingly accepts the part in the school play, and the year moves forward to graduation. Then tragedy strikes. A growing number of her PPC classmates start to "blitz." At first, it is just a matter of some students passing out in class and needing medical help, but soon one after another dies of cardiac arrest. It becomes clear that the problem is much larger. In a local news broadcast Letitia learns that as many as one-quarter of the PPCs that were engineered sixteen and seventeen years ago may be possessors of a defective chromosome sequence known as T56-WA 5659. Designed to be part of an intelligence enhancement "macrobox" used in ramping up creativity and mathematical ability, T56-WA 5659 was eventually made a standard option in virtually all preplanned children. Now the sequence is being linked to grand mal seizures in many PPCs and cardiac arrest in some.

As the story comes to an end, we learn the scope of the genetic engineering mistake. Across the nation, more than two million pre-planned children have fallen ill and a million have died. Among Letitia's classmates, this includes 200 who are sick and 112 who died. Reena Cathcart is one of the casualties. She collapses and dies in Letitia's arms during dress rehearsal for the play. Suddenly Letitia is no longer the ugly pariah but an envied "natural." Instead of enjoying her newfound status, however, Letitia grieves for her lost "brothers" and "sisters."

Asked to say something at graduation, Letitia tells her class-mates that although people made bad mistakes, the kids themselves were not mistakes. Departing from her written speech, she adds: "I mean . . . they weren't mistaken to make you. I can only dream about doing some of the things you'll do. . . . Some of you will think things I can't, and go places I won't . . . travel to see the stars. We're different in a lot of ways, but I just thought it was important to tell you . . . I love you . . . *We* love you. You are very important. Please don't forget that. And don't forget what it costs us all."

Like the many largely negative fictional visions of our genetic future, "Sisters" depicts the realization of our fears in their most terri-fying form. It poses the question of whether our quest for improved human qualities and performance will expose future generations to terrible risks. Genomic science and reproductive technologies are greatly increasing our powers to manipulate and control the human genome. But that growth in knowledge also reveals new dangers, never before imagined. The most immediate of these are physical: ways in which genetic modifications can go awry and produce harmful effects in individuals and populations. Other risks are social. By alter-ing our fundamental biology, will we disrupt healthy patterns of hu-man interaction? Beginning in this chapter with the physical risks, we shall turn in the next ones to the harms for families and society. But pointing out the risks is only part of the story. In each case, we have to ask whether we can reduce the risks to a reasonable level. Like Letitia Blakely in her grief, we should also balance risks against benefits.

Most of the methods now being employed in gene therapy research can cause the kind of harm imagined in "Sisters." Viral vectors, in particular, are prone to the kind of misplacement of gene sequences (insertional mutagenesis) that led to leukemia in three of the ten French X-SCID patients. There is a nearly universal consensus among bioethicists that these risks warrant the use of such gene therapy only at the somatic level and only as treatments for patients whose life is in peril.

Does our ability to alter DNA sequences with Mario Capecchi's precise gene targeting technique of homologous recombination change the picture? Yes and no. Yes, because achieving targeted DNA alterations that can be independently verified before they are introduced greatly reduces the risks associated with viral vectors. A 2003 report by a multidisciplinary working group of the American Academy for the Advancement of Science supports this conclusion. The report issues a stern warning against clinical implementation of inherited germline modifications unless stringent constraints are imposed. One is requiring multigenerational follow-up studies to monitor safety. But the report relaxes this requirement when dealing with the narrow category of precise gene targeting that does not involve adding novel sequences.[2]

Nevertheless, even the most targeted gene changes will not eliminate the chance of error. One problem is that most of the traits that we might want to change are not caused by single genes, as in the case of many classic Mendelian genetic disorders, but by many separated genes or gene components working together. In the language of genomics, these traits are "polygenic," or "complex." Many, if not most, genetically influenced disease conditions and traits are now believed to be of this nature. Single gene disorders or traits represent the low hanging fruit of genetic selection and modification, but complex conditions can be addressed only with finely honed genetic tools. To effectively achieve a desired result clinicians must find and precisely alter many sites on the genome.

From a technical perspective, the good news here is that genomic technologies are rapidly being developed to expedite the hunt for genetic factors in polygenic conditions. Most noteworthy is the use of gene chips and whole genome scanning for small genetic variations among individuals in large population samples. About 0.1 percent of our DNA differs from human being to human being. These differences take the form of single DNA letter changes, known as "single nucleotide polymorphisms" (SNPs; "snips" for short), that are already being identified and inscribed on high-density gene chips. Using these chips, researchers can locate many of the specific SNPs on an individual's genome. This allows researchers to correlate known features of individuals in a population with small signals in the genomic data that point, not only to single genes, but also to complex patterns of genetic causation. Using genome-wide association techniques like these, researchers have begun to identify a host of gene variations underlying heart disease, stroke, inflammatory bowel disease, and other disorders.[3] Recently, researchers used gene chip technology to examine the DNA of a group of 349 Swiss students and laborers who had previously been tested for performance in short-term memory. The study found one hundred gene variants that seemed to show up more frequently in those with superior memory ability. When the researchers repeated the experiments with two other groups, including 256 elderly people from the retirement community of Sun City, Arizona, they narrowed their findings to what they believe are five memory-related genes.[4] It will be years before these methods turn up the array of genes involved in many other diseases or nondisease traits, but what is important is that much of the science to accomplish this task is already in place.

The complexity of the genome accentuates other challenges and problems. When genomic researchers completed the first draft of the human genome in 2001, they were surprised by one finding: human beings have many fewer genes than expected. Genes are the coding regions of the genome, the DNA sequences scattered among a lot of seemingly useless or "junk" DNA. Genes produce the protein products and other factors that build the body and regulate its functions. Until the late 1990s, the presumption was that "one gene equals one protein product." Given the complexity of the human brain, and the large number of different protein products that had already been identified, researchers anticipated that humans might have as many as 100,000 different genes. Instead, they found fewer than 25,000.[5] How can scientists explain the incredible biochemical productivity of this reduced number?

The answer seems to be that the human genome is highly modular in nature. Like a language with its words and inflections, it has different component parts that can be mixed and matched to produce different meanings. The genes themselves are made up of many modular components spread physically across the genome and often separated by noncoding regions. Among these modules are sequences that produce transcription factors that determine how the DNA is read to make gene products. Also among the modules are promoter or regulatory regions that determine how much of the product is made. The existence of transcription "software" means that a gene can take many different forms, depending on how its parts are spliced together. And it can produce many different protein products, depending on how transcription takes place and what quantity of its product the cells are instructed to make. As Craig Venter and his colleagues observed when they published their draft of the genome in 2001, "A single gene may give rise to multiple transcripts, and thus multiple distinct proteins with multiple functions, by means of alternative splicing and alternative transcription initiation and termination sites."[6] Recent studies have confirmed a high rate of alternative splicing in the human genome. Between 35 and 59 percent of human genes have at least one alternative splice form, and most of these presumably have functional significance because they change the gene's protein product.[7]

Augmenting this complexity are feedback loops that modulate each component of the process. The feedback loops involve various environmental factors affecting cells and the organism as a whole. In many instances, a gene in the nucleus of a cell is not even turned on unless an appropriate set of genomic and environmental factors converge to permit it do so. As biology students have long been taught, "Genotype does not equal phenotype." That is, the particular sequence of DNA that an organism possesses (genotype) does not determine what bodily or behavioral form (phenotype) the organism will finally display. Among other things, environmental influences can cause the suppression of some gene functions and the activation of others. But our new knowledge of genomic complexity tells us that genes and parts of genes interact with other genes, as do their protein products, and the whole system is constantly being affected by internal and external environmental factors. Reflecting on these developments, Paul Silverman, one of the early leaders of the Human Genome Project, observes that in current thinking, "The gene may not be central to the phenotype at all, or at least it shares the spotlight with other influences. Environmental, tissue and cytoplasmic factors clearly dominate the phenotypic expression processes, which may, in turn, be affected by a variety of unpredictable protein-interaction events."[8] Silverman's view is not shared by all molecular biologists, who disagree about the precise roles of genes and other factors, but it signals many scientists' discomfort with a strictly deterministic view of the role of genes in an organism's functioning.

Recent research in behavioral genetics provides a fascinating illustration of these complex dynamics. As I pointed out in the last chapter, it is known that genes play a significant role in contributing to depression. Researchers have specifically identified a role for a gene (5-HTT) that affects the transmission of serotonin. People who have the so-called short form of the gene are more likely to experience anxiety and depression. Recently, however, a group of researchers led by Avshalom Caspi at King's College, London, undertook the genetic analysis of more than 847 New Zealand children whose lives were followed from early childhood until they reached twenty-six years of age.[9] As predicted, the researchers found that individuals with one or two copies of the short version of the 5-HTT gene exhibited more depressive symptoms, diagnosable depression, and suicidal tendencies than those with two copies of the long allele. But not all the subjects with the short alleles did badly. On closer analysis, Caspi and his team found that the short versions of the gene were associated with depression only in individuals who had recently (between the ages of twenty-one and twenty-six) suffered some kind of emotional trauma, such as unemployment, or a serious health or relationship problem. In those cases, environmental stress combined with the gene propensity led to depression. In contrast, individuals with copies of the long form of the gene who experienced similar stress appeared to rebound much better and throw off its effects. What this study shows (as does a very similar study by the same group on the relation between juvenile experiences of maltreatment and later tendencies toward antisocial behavior and violence) is that genes and environment work together to produce psychological effects.[10] Genes do not necessarily cause depression or violence. Particular environmental insults, including bad home environments and other stressors, may also be needed. As the science journalist Matt Ridley reminds us, "The right kind of parenting can alter an innate personality."[11] But Caspi's research shows that beneficial gene variations can moderate the response to stress, and particular gene susceptibilities can worsen it.[12]

These complexities do not make gene interventions useless. If parents can give a child a genotype more likely to produce a positive response to life's hardships, it might be worth doing so. But researchers urge caution. The importance of environmental factors suggests that some gene interventions may never evidence themselves or, in novel environments, may have unexpectedly bad effects. The modular nature and sheer complexity of the genome also suggest that, in the attempt to repair a seemingly harmful sequence, we could end up inadvertently disrupting other gene functions in ways that prove even more dangerous than the original genetic defect.[13]

The problem is compounded by information suggesting that a single human gene can have multiple effects, depending on the context of its expression in the body's cells. This phenomenon is known as pleiotropy, the contribution of a single gene to more than one trait. When it occurs, the gene codes for a product that can be used differently by different cell types. Hence, a targeted change in a gene can have beneficial side effects in many parts of the body, but it can also cause unintended harm in others. An improvement in mood can be accompanied by enhanced susceptibility to cancer.

One gene may also play very different roles at different times in life. Nature is clever and frugal. It does not always fashion new genes for new purposes. Instead, our bodies sometimes recruit older genes to accomplish complex new functions. These can be ancient genes that were long ago turned off in mammals, or human genes that do most of their work in embryonic and fetal development and then, in other species at least, are switched off for good. Recently it has been found that some of the developmental genes that shape neural development in the early-stage human embryo are turned on again later in life, perhaps after puberty, to regulate metabolic functions else-where in the body: same gene, wholly different function. As a result, even if a researcher is confident of the developmental implications of a gene modification, that confidence might not apply to the gene's role throughout the human life span.

The risks, then, seem daunting. How could we ever permit hu-man germline modifications when we are dealing with such complex systems? Won't we invite a "Sisters"-type scenario? Part of the an-swer is scientific and technical, involving better ways of understanding and managing the risks. Part is ethical. We have to think clearly about what degree of risk is permissible for each measure of harm avoided or benefit produced.

Before we go ahead with any kind of human gene modification, clinicians, researchers and oversight bodies will want to pay close at-tention to the lessons offered by the existing experiments of nature. Many of the genes related to disease eradication, disease prevention, and bodily enhancement already exist in the human population. Long-term studies of people who are insusceptible to HIV/AIDS infection because they have a mutant copy of the CCR5 receptor gene will show how safe the artificial creation of this defect by means of an AIDS gene vaccine might be. There is reason for caution here. Since most people have working CCR5 receptors, the gene probably has a useful function, although not one so essential that millions of people cannot flourish without it. It may turn out that that the receptor is redundant with other existing genes, or that it enhances response to disease condi-tions that have long been eradicated. In any case, we should probably not move ahead with widespread genetic alteration of the CCR5 re-ceptor gene until we better understand its roles and are able to make a reasoned judgment about what we gain and what we lose by disabling it in the quest for AIDS immunity. In a world of rapidly developing new biological threats, it is not yet clear whether our best bet for fighting new perils like AIDS, Severe Acute Respiratory Syndrome (SARS), or avian influenza lies with preserving the existing levels of genomic diversity or actively introducing powerful new genome modifications.[14]

As applications move from therapy and prevention to the realm of pure enhancements, the moral argument for germline-type interventions becomes less compelling and the need for justification becomes more demanding. Weighing one evil (disease) against another evil (an inadvertently introduced new genetic disorder) is never easy, but at least there are similar considerations on both sides of the scale. We are weighing apples and apples. In the case of pure enhancements, it is apples versus oranges: we are weighing a possible evil (a newly introduced genetic disorder) against benefits (better appearance, improved memory, additional years of life). Some would condemn such a calculus, saying that it is wrong to invite or impose an evil (suffering, disability, or death) merely to provide a pleasant benefit.

Avoiding serious evils is generally preferable to achieving possible benefits. We regard someone who risks life and health for the allure of a highly addictive drug as foolish. We question the intelligence of motorcycle riders who forsake helmets and expose themselves to serious head injury just to enjoy the wind in their hair. But clearly, this balancing does not always tilt against the pursuit of good. Where the risks are moderate or low, we accept them, allowing people to pursue thrills and pleasures. Downhill skiing, rock climbing, and whitewater rafting are all sports that depend on this assumption. Enhancements also flourish in the medical realm, even when they entail a small risk of serious harm. People undergo cosmetic plastic surgery for minor appearance problems, even when the surgery carries a small risk of disfigurement or, in rare cases, death.

A special problem for germline genetic enhancement is that

those who invite the risks are different from those who suffer them. Parents authorize these interventions, but if things go wrong, their children suffer the consequences. An adult can choose to risk his life climbing mountains. But does he have a right to similarly risk his child's life? In fact, parents frequently make decisions that jeopardize their children's welfare and physical safety. Generations of missionaries have taken their children to remote and dangerous parts of the world. Affluent parents take children along on potentially risky adventure trips. And people routinely encourage their children to participate in school football, hockey, or lacrosse, sports that carry known risks of serious physical injury. It is hard to understand why we permit this behavior, unless we are prepared to acknowledge that parents' aspirations for themselves and their children sometimes take precedence over a child's physical security.

Where do these distinctions between avoiding harm and seeking benefits and between inviting and suffering risks lead? I think they tell us, first, that the requirement of evidence for the long-term safety of any genetic intervention grows proportionately as we move from somatic measures to germline interventions. It increases further as we move from treatments and preventions to pure enhancements. In each case, greater risks are justified only if the harm avoided or benefit obtained has a greater magnitude than the risk invited. There is a far more compelling case for attempting to develop an AIDS gene vaccine than a cosmetic enhancement that whitens tooth enamel. On the other hand, if the dental enhancement is demonstrably of low risk (perhaps because there is already a safe natural variant), then the risk-benefit calculus tips back in its favor.

Genetic interventions that pose the threat of a catastrophic risk like the T56-WA 5659 gene in the "Sisters" scenario are the greatest source of worry. But new technologies may lend a hand. One of these is new computer-based scanning of gene functioning in cell cultures.

Affymetrix, the gene chip maker mentioned at the start of this book, is a leader in this technology. Affymetrix's silicon DNA microarrays, each no larger than a postage stamp, contain millions of DNA sequences.[15] When the chips are bathed in a cell's fluids and scanned by laser light, they can detect sequences active in the cell at any given moment that bind to the chip's template. Affymetrix's DNA chips are being used to scan the whole human genome for gene combinations associated with cancer. In the future, they will be used as a testing ground for any intended gene modification before it is inserted into a living organism; the chips will tell how that modification is likely to interact with other parts of the genome. Similar technologies used at the level of the genes' products, proteins, in a new science known as proteomics, are enhancing our ability to understand what actually goes on inside living cells.[16] Given the current pace of progress, within a decade or two, the expression of every DNA sequence in every cell and tissue of the body will likely be in a database and can be anticipated by those making genetic changes.

One of the most important areas of technological advance concerns the possible *reversibility* of genetic changes. Earlier I mentioned Mario Capecchi's work in connection with homologous recombination. But another aspect of Capecchi's research has great importance for the field of gene therapy—his invention of a method for switching gene modifications on or off. Capecchi and his colleagues have identified an enzyme, CRE recombinase, that has an unusual property. When activated by a special drug, it deletes all DNA letters that it finds between two DNA sequences called LoxP sites. These sites are thirty-four nucleotides in length. By chemically stitching a LoxP site to both ends of any added gene sequence and activating CRE recombinase, Capecchi's team was able to delete the new DNA sequence at will, much as a word processor is able to find and delete an undesired sequence of letters. In a series of experiments, Capecchi's team produced a line of mice with germline modifications bracketed by LoxP sites. Then, in the first generation of mice, they deleted the added sequence, entirely erasing the new modification from the lineage.

In further research, Capecchi's team was able to use the CRE recombinase–LoxP approach to *reintroduce* a sequence that had previously been deleted. They did this by exploiting another feature of the LoxP sites. When two LoxP sites are inserted into the genome with their own strands of DNA letters lined up in opposing order and are then exposed to CRE recombinase, the sites do not delete but *invert* the order of any sequence between them. Transcription machinery usually reads each strand of the DNA double helix in a specific direction. Using this technique, therefore, a nonfunctional inverted sequence (GCAT) can be restored to its functional original form (TACG). In one experiment, the researchers replaced an existing gene with a new one. But they also left in place an inverted copy of the original sequence. By properly positioning LoxP sites around the various sequences, they were able to use CRE recombinase to delete the replacement sequence and invert and reactivate the altered gene. Like a writer who had retained a hidden-text backup copy of a revised sentence, they were able to erase the changes and recover the original.[17]

This research has several implications for gene therapy safety. First, it tells us how we might avoid the "Sisters" scenario. We could reduce the risk of genes going haywire if, at least during the early years of gene manipulation, we required that all human gene modifications be made reversible. While this might not eliminate all risks (some errors might do lasting developmental damage before they could be reversed), it would limit them. It would also give each subsequent generation the option of deciding not to pass inherited changes on to their children. We could also use these switching techniques to narrow the scope of the changes we introduce. For example, we might be interested in altering the expression in adulthood of a gene that also has multiple consequences during fetal development. To prevent developmental impacts, we could alter it only later in life by administering the appropriate drug at that time. One illustration of this approach is based on the recent finding that certain genes in the body contribute to aging by rapidly destroying a cell if it shows any signs of malfunctioning caused by oxidative stress.[18] Destruction of malfunctioning cells is one of the body's first lines of defense against cancer. In a younger person with many viable cells this mechanism makes sense, since cancer is a greater risk than the loss of cells. But in an aging person, the process is costly and contributes to the tissue loss that hastens disability and death. By inserting gene variants that slightly dampen the hairtrigger mechanism that starts this process, and by keeping those genes quiescent until late in life, researchers could develop an effective and safe way of maintaining bodily vigor throughout the life span. On a lighter note, we could use the same approach to repair the genes that contribute to premature hair loss and baldness (alopecia). Why should we expose someone to altered genes throughout his life (and perhaps the risk of excessive bodily hair) when we could repair sequences that are kept switched off until hair loss begins? With the first signs of a receding hairline the person could apply a topical cream to activate the CRE recombinase and rejuvenate the hair follicles in the scalp—and only those hair follicles.

The health risks that human gene modifications could impose on our children and their descendants are very serious. Because they place identifiable people in peril, they take precedence over some of the more speculative concerns that we shall look at next. These health risks demand caution, but they do not necessarily bar modifications in the future. We can already see some of the scientific breakthroughs that may lower the risks, and we can envision some of the steps ahead. At the start, successful gene therapies for disease conditions and effective gene-based preventive measures will increase our knowledge of how to implement this technology and our confidence in its safety. Following that, we may step across the line to pure enhancements, relying on well-researched natural variations already present in the population as a measure of safety. Insisting, whenever possible, on the reversibility of changes will help increase the margin of safety. Occasional mishaps may well occur, but if we proceed cautiously and if we are lucky, the tragedy of "Sisters" will remain science fiction.

The ability of gene therapies to eradicate an inherited disorder like sickle cell disease raises a question. Apart from the immediate dangers of gene interventions gone awry, could altering harmful genes in the population have bad consequences for the human species as a whole? Could the process eventually lead to a reduction of genetic diversity, whose loss we might regret if new pathogens or adverse environmental conditions appear?

Sickle cell offers a textbook illustration of how, in appropriate circumstances, even a harmful gene mutation can prove beneficial. The disease exhibits the phenomenon known as "heterozygote advantage." Heterozygotes have only one copy of a mutant gene. Since sickle cell is a recessive disorder, heterozygous people carry the sickle cell trait but do not usually get sick. It is now known that the mutation that causes sickle cell arose several thousand years ago in parts of Africa where the parasite that causes malaria had recently made its appearance. Over time, the mutation became widely disseminated in those populations because, even though people who inherited two copies of the gene often died, those with just one copy, the heterozygotes, had significantly improved resistance to malaria. As a Punnett Square will show, for a recessive Mendelian disorder like this there are, on average, two heterozygotes born for every affected homozygote (plus one unaffected child without the mutant gene). Thus, overall, these families produce three healthy offspring for every sick one.

Heterozygote advantage explains why some harmful mutations persist and proliferate. Although the mutations are reproductively disadvantageous and even lethal for the minority who inherit two copies of the mutant gene, the slightly improved reproductive success of the many heterozygotes carrying the mutant gene spreads the mutation in the breeding population. It is now believed that the same dynamic explains the presence of a host of other recessive genetic disorders. For example, in its heterozygous form, cystic fibrosis is believed to provide significant protection against cholera.[19] The genetic mutation may have spread among northern European populations during periods when the growth of towns and cities created the environment for cholera epidemics. The result is that in Great Britain today, 1 in every 24 people is a cystic fibrosis carrier, and 1 in every 2,400 children is born with the disease. Similarly, for Tay-Sachs disease, a neurodegenerative disorder that is highly prevalent in eastern European Jewish populations and leads to early childhood death, some have theorized that carrier status may protect against tuberculosis.[20] Tuberculosis was epidemic in the dense ghetto environments in which many Jews were forced to live. In all these cases, a small number of people suffer grievously, and the disease resistance of the rest of the breeding population is slightly enhanced.

Considering these examples, is it wise to modify genes today, even markedly harmful genes, if they may have unidentified health benefits, or if, in the event of future epidemics or catastrophes, they may in some way prove to be beneficial? Stating the question this way shows how morally dubious it is. Would anyone tell the parent of a child suffering from a severe case of sickle cell disease, "I'm sorry, but we're not going to use this effective new gene therapy to cure your child because it may influence the future prevalence of alleles that resist malaria"? Quite apart from the fact that many of the populations that carry this allele are no longer exposed to malaria, the trade-off here is morally offensive. We do not usually ask people to suffer illness just to preserve a beneficial genetic population for others. Instead, we try to prevent or cure the inherited disease while taking other public health or medical measures to reduce its prevalence—in this case, by efforts aimed at eradicating malaria. Where the condition is undesirable, we privilege its elimination over worries about remote genetic consequences. This logic would certainly apply to germline genetic modifications aimed at preventing disease, even if these happened to eliminate a beneficial heterozygote effect of a harmful sequence.

The same reasoning applies to a different class of genetically influenced conditions: those whose eradication may bring about the loss of other beneficial human talents and abilities. The case most often mentioned in this respect is manic depression, or bipolar disorder. This illness tends to run in families, and researchers are now focusing on several genes that are believed to significantly contribute to the disease. In severe cases, the episodes of manic energy, marked by racing thoughts, grandiose thinking, and spending sprees, alternate with periods of sadness or bleak despair sometimes leading to suicide. The illness can ravage a person's life and tear apart family relationships. But manic depression is not altogether without its positive aspects. Although lithium therapy dramatically reduces the symptoms of the disease in many patients, some sufferers resist taking the medication because they find the roller coaster of moods exhilarating, or because they depend on the manic spells for their creative energies. The work of Kay Redfield Jamison has shown that many writers and artists appear to have suffered from the disease.[21] The association of manic depression with creativity raises the question of what society will lose if we erase this particular genetic propensity from our midst by gene selection or modification. Will we end up with fewer people suffering severe emotional problems but also with less artistic creativity and innovation?

These are real concerns. But in the case of manic depression, I can pose the same questions that I asked in connection with sickle cell disease. Would you inflict severe manic-depressive disorder on someone just to preserve the possibility that he or she could turn out to write poetry at the level of William Blake or Sylvia Plath? What about the many other sufferers for whom this disease produces not artistic achievements but bankruptcy as they gamble away family resources or risk their own and others' lives?

In thinking about these choices, we should also keep in mind the advancing state of genomic knowledge. In June 2005, I attended an NIH workshop that focused on new techniques for whole genome sequencing.[22] This was in connection with the thirteen-year, three-billion-dollar project to identify the order of DNA letters in the genome.[23] The workshop focused on the new sequencing machines and technologies being developed, which by the year 2011 could reduce the cost of identifying a person's genome more than a thousandfold. This is the $1,000 genome I mentioned earlier. When it comes about, medicine will become personalized: the sequencing of anyone's genome will be a routine part of his or her medical care. With your whole genome in hand and a growing database of sequences known to be associated with a higher risk of disease, your doctor could take steps to protect you from disease many years before it occurred. For example, if your sequence showed you to be at risk for colon cancer, you could start early, perhaps in your teens, on a diet likely to reduce the occurrence of the disease, or you could begin an intensified screening program for colon polyps.

The $1,000 genome is also an answer to our question about genetic diversity. Although gene interventions may reduce the incidence of some genes in the human population, genomic technologies are now making it possible to detect, measure, and record that diversity as never before. This means that while we might erase the sickle cell mutation from the genomes of most living people, it will still be possible to keep that trait alive in the form of computerized information. In the future, data banks will contain this sequence as well as the complex medical histories of its consequences. If we ever need to recover features of this sequence, say, for antimalarial gene therapies, we could do so directly from the information resources rather than from living people. The key insight here is that DNA sequences, before they become biological realities, are information. As the science writer Matt Ridley points out, "Genes are just chunks of software that can run on any system."[24] What we are now doing is learning how to read, store, and, where necessary, reinsert that information into living individuals or animal models for biological expression.

The upshot is that those opposed to gene modifications on the grounds of population effects might be better advised to support efforts to accumulate and preserve the full record of human genetic diversity. These include the Human Genome Diversity Project, an effort to collect samples of human DNA from around the world.[25] This project, which began in the early 1990s, is currently halted because of concerns over patenting and the commercial exploitation of DNA donors. If those concerns could be resolved, we might be able to develop an invaluable genomic snapshot of the human species at this moment in history. In mid-2005, National Geographic and IBM teamed up to sponsor the Genographic Project, an effort to collect at least one hundred thousand DNA samples from people all over the globe to trace the routes of human migration.[26] The availability of information like this tells us that the population-genetic risks of gene modifications are real but manageable. Increasing genetic knowledge will amplify our powers to understand, record, and, if necessary, recover human genetic diversity.

For humans to flourish depends on much more than biological survival and the ability to reproduce. We are social and cultural creatures. This means that we have to look at modifications of the human genome in terms of impact on the ways we live together. In the previous chapter, we saw that extended life span is one of the enhancements that parents might want to offer to their children. But what is good for individuals is not always good for society. Sometimes technological innovations have unanticipated social side effects. The destruction of fisheries by factory fleets and the buildup of greenhouse gases through the excessive use of fossil fuels are examples. Things that are tolerable when they are scarce become harmful when overused. The widespread extension of the human life span is one genetic enhancement that provokes considerable worry.

In October 2003, the President's Council on Bioethics, the Bush administration's bioethics committee, published the report entitled *Beyond Therapy* that assessed a host of pharmaceutical and genetic technologies aimed at enhancing normal human abilities. The report devotes an entire chapter, "Ageless Bodies," to the issues of age retardation and life extension. It distinguishes between the two, defining age retardation as the effort to slow the onset of the physical and mental degeneration associated with aging in order to ensure a healthy old age within the existing limits of human life, and life extension as the effort to prolong the absolute duration of life. The council acknowledges that the objective in either case is not just a prolonged old age if that merely extends the period of decline and dependency. The goal is to enhance functioning throughout all the stages of life, especially the final ones: "not only to add years to life, but also to add life to years."[27]

After outlining some of the ways that genetic research could accomplish both these purposes, the council explores in its report how such changes might affect society and the quality of human life. On the positive side, it notes very briefly an intrinsic moral benefit: "A significantly greater life span would open up new possibilities and freedoms." Society, too, may benefit, "gaining much from the added experience and wisdom of its older members."[28]

But in the council's thinking, the negatives predominate. For example, the report asks whether postponing both aging and dying might not adversely affect the way individuals approach the tasks and challenges of life. "Many of our greatest accomplishments are pushed along, if only subtly and implicitly, by the spur of our finitude and the sense of having only a limited time. A far more distant horizon, a sense of essentially limitless time, might leave us less inclined to act with urgency."[29] The result could be a loss of social energy and creativity, "a life of lesser engagements and weakened commitments."[30]

The report expresses worries about the impacts on family. If children are our answer to mortality, would enhanced longevity weaken the urge to reproduce? The council believes so. "A world of men and women who do not hear the biological clock ticking or do not feel the approach of their own decline might have far less interest in bearing—and, more importantly, caring for—children."[31] And what of matrimony? "Would people in a world affected by age-retardation be more or less inclined to swear lifelong fidelity 'until death do us part,' if their life expectancy at the time of marriage were eighty or a hundred more years, rather than, as today, fifty?"[32]

Above all, the council worries about the social and economic conflicts likely to be precipitated by a rapidly expanding, but still active older population: "Sons might no longer surpass their fathers in vigor just as they prepared to become fathers themselves. The mature generation would have no obvious reason to make way for the next as the years passed, if its peak became a plateau. The succession of generations could be obstructed by a glut of the able. The old might think less of preparing their replacements, and the young could see before them only layers of their elders blocking the path, and no great reason to hurry in building families or careers—remaining functionally immature 'young adults' for decades, neither willing nor able to step into the shoes of their mothers and fathers."[33]

I offer these observations by the president's council not so much to present an exhaustive evaluation of the social and psychological implications of life extension technology as to suggest that we are probably not well equipped to think about such basic biological changes in human life. Trying to assess the possible social implications of any major change, whether it be improved overall appearance, increased IQ, or greater longevity, asks us to extrapolate from our current knowledge and values to circumstances that are likely to be very different because of those changes. Faced with the difficulty of imagining or valuing change, it is very easy to fall back on what I earlier called status quo bias. People thinking this way privilege the present and feel threatened, sometimes irrationally, by any departure from it.

This is immediately evident when we consider the dramatic extension in the human life span in the past hundred years. As the report acknowledges, life expectancy at birth went from about forty-eight years in 1900 to seventy-eight years in 1999, an average increase of about three years per decade.[34] These numbers are slightly skewed by effects of dramatic declines in infant mortality through public health programs and better nutrition. But if we control for this by taking the statistic of average life expectancy at age ten, a measure that takes infant mortality out of the picture, we still see striking progress. In 1900, if a white female managed to reach age ten, she could count on an additional fifty-two years of life, dying on average at age sixty-two. By 2004, the same white female at age ten could count on an additional seventy-one years of life, a 31 percent increase in longevity. The averages are somewhat lower but also greatly improved for white males and nonwhite males and females.[35]

Does anyone think that these changes have been unfortunate? Although it is true that the large number of baby boomers reaching retirement is causing Americans to worry about viability of the current Social Security system, no one is calling for an end to the public health and medical programs that have brought about this enviable situation.

Recent reports even suggest that the improved health status of many older people today may lower medical costs for this generation of retirees.[36] Imagine how a presidential council in 1900 might have regarded the 31 percent increase in life span that was going to occur in the century ahead. Would it also have worried about the potentially devastating impacts on the economy, family, and generational relationhips? Would not all such worries, like the 2003 council's report, reflect status quo bias rather than reasoned reflection?

The Oxford philosophers Nick Bostrom and Toby Ord, in the course of defending the possibility of human biological enhancement (including genetic enhancement), have offered a conceptual test for determining when status quo bias is present in our thinking. They call this the "reversal test": "When a proposal to change a certain parameter is thought to have bad overall consequences, consider a change to the same parameter in the opposite direction. If this is also thought to have bad overall consequences, then the onus is on those who reach these conclusions to explain why our position cannot be improved through changes to this parameter. If they are unable to do so, then we have reason to suspect that they suffer from status quo bias."[37]

In the case of age extension, the reversal test asks us to look at our present life span and consider not only whether we would be prepared to extend it by, say, 31 percent but whether we would also be willing to abbreviate it by the same amount. If the answer to the first question is uncertain and filled with worries but the answer to the second is a firm "No!" then the test tells us that status quo bias may be influencing our thinking. For no clear reason, we have concluded that in terms of life span we live in the best of all possible worlds. But why are we worried about the future when the same types of changes in the past were welcome? Is it just that we have grown used to the present and irrationally fear the future? The reversal test checks for status quo bias by asking us to closely examine the validity of each of our reasons for worrying about change. If we are not willing to seriously apply those reasons to our current state, the reversal test tells us that they may not be valid reasons at all but simply expressions of fear.

I do not mean to use this argument to dismiss all the concerns that people have about the social impacts of genetic enhancement technologies. There are justified reasons for worrying about the impact of some of these changes on families. There are also concerns that genetic technologies could introduce entirely new dynamics into our age-old problems of social justice and could greatly widen the gap between the rich and the poor. The disturbing history of the eugenics movement shows that there are good reasons to fear the social impact of a misguided quest for genetic improvement.

But status quo bias permeates much of the thinking about genetic technologies. Like the German villagers mentioned in the introduction who favored their purely accidental village layout over any alternative, we tend to accept our familiar reality without subjecting it to the kind of critical scrutiny that we apply to changes. In this sense, the present, no matter what its faults, is always the best of all possible worlds. The current human life span is perfectly acceptable, whereas age extension raises frightening possibilities. The present range of human IQs is fine, but improvements look dangerous. The existing range of human heights and facial attractiveness is good, but the pursuit of enhanced appearance raises concerns about "harmful conceptions" of beauty.[38] In each of these cases, the reversal test forces us to ask such questions as "Would it be better if we reduced average human IQ today across the board by fifteen points?" and "Would it be better if we banned reconstructive surgery for cleft palate, or cosmetic rhinoplasty for people who are troubled by their noses?" If the answer to these questions is no, we have to ask why we find the status quo satisfactory but modest improvements so worrisome.

Bostrom and Ord's reversal test is a generally useful caution as we go about evaluating biomedical innovations, especially the prospects for the genetic modification of our offspring. Because these changes represent a radical departure from the way we approach reproduction, it is easy to conjure up fears. In the next chapter, we shall look at concerns that have been voiced about what the technology of genetic enhancement might do to parenting. Here the limited number of ways that parents currently shape their children's bodies and lives provide the basis for a test. These include such things as orthodonture for more attractive teeth, drugs to control attention deficit hyperactivity disorder, and supplemental education services to overcome learning problems. Would we halt or reverse these interventions if we could? If the answer is no, then we must show why we would reject similar, genetically based efforts in the future.

CHAPTER 5

Parents: Guardians or Gardeners?

A squat grey building of only thirty-four stories. Over the main entrance the words, CENTRAL LONDON HATCHERY AND CONDITIONING CENTRE, and, in a shield, the World State's motto, COMMUNITY, IDENTITY, STABILITY.

The enormous room on the ground floor faced towards the north. Cold for all the summer beyond the panes, for all the tropical heat of the room itself, a harsh thin light glared through the windows, hungrily seeking some draped lay figure, some pallid shape of academic goose-flesh, but finding only the glass and nickel and bleakly shining porcelain of a laboratory. Wintriness responded to wintriness. The overalls of the workers were white, their hands gloved with a pale corpse-coloured rubber. The light was frozen, dead, a ghost. Only from the yellow barrels of the microscopes did it borrow a certain rich and living substance, lying along the polished tubes like butter, streak after luscious streak in long recession down the work tables.

"And this," said the Director opening the door, "is the Fertilizing Room."[1]

The beginning of Aldous Huxley's 1932 novel *Brave New World* depicts a society where normal human families and parenting have been replaced by industrial manufacture. This opening passage, with its glimpse of the Central London Hatchery and Conditioning Centre, presents a deathly image. Human conception and birth have been drained of all vitality and replaced by cold, mechanical processes. In the hatchery's interior, assembly lines fertilize hundreds of human eggs. The resulting embryos are nurtured in glass canisters until birth. Using cloning technology, genetic and chemical manipulations, and post-birth psychological conditioning, the centre produces a docile population well suited to its totalitarian state. Children emerge from the centre as Alphas, Betas, Gammas, or Epsilons. Alphas are at the top of the genetic hierarchy. They rule society, and the lower-ranked and genetically deformed Betas, Gammas, and Epsilons perform servile tasks. But all the products of the centre, regardless of their place in this manufactured hierarchy, are programmed to believe that their repressive and emotionally sterile society is perfect.

Brave New World reflects its era. In Russia and Germany, dictators were experimenting with replacing the family with state-run nurseries and thinking about applying methods of animal breeding to human populations. In the United States, thanks to the introduction of industrial methods of mass production, millions of identical products, from tubes of toothpaste to automobiles, were rolling off new assembly lines. It was natural to ask, "If we can bring such efficiency and perfection to industrial manufacture, why not to human reproduction?" Huxley's brave new world seemed a logical end point for the application of technology to human life.

Indeed, *Brave New World* captures fears about human genetic self-modification that have troubled people for more than seventy years. The very mention of human genetic manipulations evokes Huxley's stark vision. Part of the problem is political: Huxley's is a totalitarian society. By replacing the family, the state is able to produce perfectly obedient citizens who lack the ability to resist its demands. But the fear stirred by the novel goes much deeper. It is the fear that new forms of reproduction will replace the family and, in doing so, eradicate everything that makes life worth living. The family is the symbol and source of many cherished hopes: for love, respect, and commitment.

The family may be the center of human life, but the industrial and technological society that forms the background to *Brave New World* magnifies the family's importance as our *emotional* center. The novel remains powerful because all around us the world has grown cold and impersonal. Today most people in the developed world live not in small towns or villages, surrounded by friends and kin, but in urbanized areas. We no longer work in the home or on the farm but in large organizations, where others often know us as nothing more than a phone number or E-mail address. From infancy and throughout years of schooling and employment, we are measured, evaluated, sorted, and stamped for acceptance or rejection according to performance criteria. Those who fail to live up to specification are ejected from the social assembly line and consigned to society's lowest ranks. Against all this pressure, the family is widely seen as our last and best line of defense. We like to think of it as the one place where we love others unconditionally and are unconditionally loved by them, the place where imperfection and failure are tolerated, where the weak are supported, and where lifelong acceptance and commitment prevail over evaluation and selection. The family is where we retreat for human intimacy. We sense that even if the new reproductive technologies promise to benefit us by reducing the burden of disease or improving human performance, we must fight those technologies if they threaten the family.

Critics of gene modification have offered four major reasons why they believe that expanded programs of prenatal genetic choice could harm both children and the family. First of all, they fear that if parents are able to determine their children's inherited characteristics, healthy parental love will be replaced by critical evaluation. Parents' response to a new child might go from excited appreciation to asking how well the newcomer measures up to expectations. In its report *Beyond Therapy,* the President's Council on Bioethics expresses the concern:

"The attitude of parents toward their child may be quietly shifted from unconditional acceptance to critical scrutiny: the very first act of parenting now becomes not the unreserved welcoming of an arriving child, but the judging of his or her fitness, while still an embryo, to become their child, all by the standards of contemporary genetic screening."[2]

Changed parental expectations can have a direct effect on the child's life. Will parents neglect a child that doesn't display the traits they ordered? Will they reject their offspring, putting the child up for adoption or using foster care to turn their "damaged goods" over to society? In the effort to fulfill their expectations, parents may force the child into a mold. Will parents constantly pressure a child genetically engineered for ability as an ice skater into a life of athletic training?

A second major reason why people worry about expanded programs of prenatal choice has to do with the engineered person's psychological sense of freedom and self-esteem. When a child has been given a set of genetically influenced talents, abilities, or even interests, can the child take credit for personal accomplishments, or must the child regard them as the work of the parents and their consultant geneticist? Are the child's failures the result of programming errors, or are they something the child has personally done (through laziness or neglect) to mess up an otherwise good genetic endowment? In short, do genetic selection and modification put the child in a lose-lose situation, where all accomplishments are chalked up to others and all shortcomings are personal failures?

Third, people worry about the impact of genetic manipulations on our right to make our own life choices. Do parents' choices of our genetic traits reduce our ability to shape our future as we wish? Does genetic selection give parents a power over us that no human being has ever possessed before: the power to shape our very nature and direct our choices? Apart from the social and psychological effects of genetic interventions, do interventions ethically violate what the philosopher Joel Feinberg calls the "child's right to an open future"?[3]

Finally, people fear that genetic choices will morally deform the way parents should treat their children and, through them, people in general. By making one or another valuable characteristic the reason to bring someone into being, don't we confuse the person with the trait? Doesn't a culture of prenatal genetic selection erode our respect for persons by mistaking the part for the whole and making superficial qualities the "measure of the man"? We like to think that human beings deserve respect in and of themselves, but trait selection seems to reduce the complexity of a person to surface qualities. It may also insert disrespect for others into the heart of family life, the place where we expect children to learn how to treat other people properly.

These four concerns do not exhaust the list of worries about the familial impact of genetic interventions. I emphasize them because I take them so seriously. In many ways, they make intuitive sense. If I create a child on the basis of my expectations, how can I not expose myself and the child to the worrisome consequences of failure? How can a mentality of expectations not impair the unconditional love that should be at the heart of the parent-child relationship? How can a culture that bases relationships on superficial trait selection not demean human beings?

But sometimes, ideas that seem obvious are not right. Reality is more complicated than that. We miss the fact that parental love is powerful enough to provide a corrective to some of these fears. In addition, status quo bias in the form of uncritical acceptance of current practices leads us to ignore serious imperfections in how we currently conduct our reproductive and family lives. When we examine these seemingly obvious worries more closely and measure them accurately against present realities, they are less impressive. We can even begin to imagine a future in which genetic modifications of children may *improve* the circumstances of families and children. In what follows, I do not intend to dismiss these four concerns. Together, they constitute serious objections to gene modification. But I hope to start us thinking critically about our anxieties and even to suggest some ways that we can increase the chances that gene modification will improve, not threaten, family life.

We think of the family as an organic environment, where relationships grow naturally out of our most basic human instincts and emotions. Gene selection and modification threaten to replace this natural environment with images of engineering design, manufacture, and consumer choice. As the President's Council on Bioethics states, "With genetic screening, procreation begins to take on certain aspects of the *idea*—if not the practice—of manufacture, the making of a product to a specified standard."[4] Others fear the turn toward a market mentality. In the words of a report issued by the Washington, D.C.-based Genetics and Public Policy Center, "Some believe germline genetic modification and other reproductive technologies will change the nature of the love parents have for their children by making children a commodity that parents have produced to their specifications rather than a gift to be loved 'to the point of irrationality.'"[5]

We do not have to rely only on our intuition to imagine these changes. Fiction writers can help. In a series of novels written between 1993 and 1996 known as the *Beggars* trilogy, the author Nancy Kress follows the lives of a group of genetically modified people over several generations. One of these is Leisha Camden. At the start of *Beggars in Spain*, the first novel in the trilogy, we witness Leisha's conception and birth. Her father, Roger Camden, is a successful, self-made Chicago businessman who learns of a program at the nearby Biotech Institute to produce children with enhanced capabilities: better looks and higher IQs. He also learns of a modification secretly under testing by the institute that allows children to go without sleep. Roger is intrigued because he is so hard driven himself and functions well on less than four hours' sleep a night. He believes that the modification will further enhance his child's abilities. Using legal threats, Roger persuades the institute's director, Dr. Ong, to agree that the first child produced for him and his wife Elizabeth will have the "Sleepless" modification.

Everything seems to go as planned. Leisha is as remarkable a child as Roger had hoped. Bright, healthy, and joyous, she is the apple of her father's eye. Unfortunately, during her conception, a second egg was present in Elizabeth's womb that was unintentionally fertilized by Roger's sperm. The result is an unmodified, natural "Sleeper" child named Alice. From birth, Alice is spurned by her father and relegated to the care of a nursemaid. Elizabeth tries to provide Alice with emotional support, but under relentless pressure from her husband's demands, Elizabeth sinks into alcoholism. Roger eventually divorces her and marries Susan Melling, the genetic scientist who helped produce Leisha. Thus the organic family of mother, father, and children is replaced by an artificial techno-family committed to an ethic of perfection. Over the sequence of novels in the trilogy we follow Leisha as she grows to brilliant maturity and Alice as she wanders off into a series of tragic misadventures. Eventually Leisha's cohort of Sleepless children offers society a growing array of scientific and material achievements—and spurs disturbing social turmoil.

Beggars is a kind of fictional controlled experiment. Roger Camden's two daughters, Leisha and Alice, one genetically enhanced, one natural, illustrate our fear that genetic interventions and expectations will not only harden parents' attitudes but also endanger their children.

How can it be otherwise? If I choose or design a child with certain traits, won't I feel pleased if I succeed and disappointed if the child fails to live up to specification? I may not give my failed product up for adoption, but won't our relationship be tarnished by my sense that this is not what I ordered? A consumer metaphor is instructive here. If I am attracted to a silver Ford Explorer with black leather seats, and I order one from my dealer, I expect to get that car. What happens if, after several weeks' wait, the salesman informs me that the vehicle is no longer available in that color, that all I can get is a blue model with a beige interior? I may accept it. But something is wrong from the start. I do not quite bond with it. Over time, I may neglect the car's maintenance and even trade it in early for another model. We can tolerate such dissonance and neglect where automobiles are concerned. But if we convert parenting into the choice of a consumer product, won't we dangerously expose all of our children to these risks?

As compelling as these ideas are, they are misleading. The parent-child relationship not only should not be equated to an owner-car relationship but is different in psychological terms. Arranging for gene modifications in one's child is not like ordering a custom car. There is a dynamic in parenting that the fears miss. I raise this dynamic to the level of a psychological principle that, somewhat whimsically, I term PLAAP: "Parental Love Almost Always Prevails." This principle sets forth the common truth that most parents end up loving whatever child they receive, no matter how much he or she conforms to—or disappoints—their prebirth expectations.

A small example: My daughter very much wanted a girl for her first child. She felt she could better understand and parent a girl, but she had a boy. In the five years since my grandson's birth, my daughter has told me many times how thrilled she is with her son, that she could not imagine this wonderful child to be other than what he is. Here is PLAAP in action. Parents love what PLAAPs into their laps.

Exceptions to this psychological rule must exist, but almost none of them have to do with failed expectations. Psychological studies suggest that when a parent fails to bond to a child, other factors in the parents' lives are usually at work.[6] The parent may be immature or otherwise unready for a child, or the marriage may be troubled in ways that disturb the parent-child relationship. One or both parents may suffer from serious psychological problems (perhaps a history of abuse). The child may have very serious health or developmental problems that overburden the parent and interfere with parental love. But disappointing prebirth expectations is rarely a factor in these problems.

Interestingly, debates about the impact of genetic screening on people with disabilities furnish some evidence for the power of PLAAP. Defenders of the rights of people with disabilities have expressed concern that the increasing use of prenatal genetic screening will damage communities' acceptance of disabled people. They fear that efforts to avoid the birth of children with serious congenital problems will make us less willing to have disabled people in our midst. They ask whether parents who are led to expect flawless children might be unwilling to raise the child who slips through the screening net.

In response to these concerns, some disability rights advocates have argued for limiting people's access to prenatal screening.[7] Others, less willing to interfere with parental choice, have advocated that we introduce mandatory educational programs to accompany prenatal screening. They argue that the public has unreasonable fears of disabilities and tends to magnify the hardships that accompany them. Some recommend that parents who learn after testing that they are likely to have a child with a disability should be asked—or even required—to meet with parents raising a child with the same condition.[8] What these people will learn, the disability rights advocates insist, is that parents love their disabled children, often regard them as the highlight of their lives, and usually adjust their various family responsibilities to meet the extra needs of their "special" child. Indeed, they point out that parents of children with disabilities are among the most vocal critics of widespread genetic prenatal screening.

What can we say about these arguments? First, I think that they must be taken as only a part of the truth. Parents who adjust best to the reality of having a disabled child are among those most likely to want to share that experience with others. In the background are other families for whom the burdens of raising such a child prove daunting. None of these arguments, therefore, should be taken as a basis for rejecting prenatal screening or encumbering it with "educational" requirements that block people's access to screening. But these parental reports should also calm our fears about prenatal genetic interventions. What they demonstrate is how powerful parental love—and the PLAAP principle—is. Parents almost always bond to children *as they are.* Even children with the most severe disabilities are usually accepted and loved. Parents will frequently say that they can't imagine the child without his or her special needs and that these needs enter into their love for the child.

So it is not clear to me that genetic testing or gene enhancement will endanger parental love. Parents will try to produce the children they desire, but in almost all cases, they will love the children they get no matter what qualities they possess. Parents who wish for abled children will accept and love disabled ones; parents who try to have enhanced children will accept and love average or disabled ones. Those who think otherwise and believe that gene interventions imperil children do so because they mistakenly apply the culture of consumerism and manufacture to the parent-child relationship. In doing so, they overlook the power of the PLAAP principle.

Roger Camden's rejection of his unenhanced daughter, Alice, epitomizes the kind of immediate harm that many believe trait selection could visit on a child. Although this fictional example may be extreme, people worry that even when a parent emotionally accepts a genetically modified child, the child will be subject to lifelong pressure to conform to the parent's expectations. If I choose traits for my child and expend effort and money to see that he or she has them, won't I do everything possible to ensure that those expectations are fulfilled? Won't I do this even if it runs counter to the child's own interests and leanings? The problem grows when we realize that gene manipulations will probably never be perfect. Each of us is the result of various genetic and environmental factors, all of which are further affected by the range of choices that we make. Even if we assume that genetics played some role in Albert Einstein's intellectual creativity, the cognitive skills that led him to accomplish breakthrough physics in the early twentieth century might today lead a similar child into a passion for rock music. If I force my budding Einstein into a science career, not only might I harm him but I could deprive the world of his unique creative contributions.

What about parental pressure like this? First, we should note that overbearing parents are not confined to the world of prenatal genetic interventions. Parents already bring considerable pressure to bear on their children to develop talents or hone skills that the parents cherish or perceive in their offspring. They struggle to enter their children into the best preschools, the best grade schools and high schools, and the best universities—all in the effort to realize their own dreams of the child's accomplishments. How many times have I attended reunions of Dartmouth classes to find tiny infants decked out in T-shirts with words like "Class of 2024" inscribed on them? We smile, but the seriousness of this intent and the pressure that accompanies it become evident when a decade or two later a child is not accepted at Dartmouth. The extent of parental disappointment reveals itself when alumni giving abruptly ends. The child, too, may feel responsible for "letting Mom and Dad down."

Of course, pressure like this does not represent the best in parenting. Won't gene interventions only intensify the pressure and give parents new reasons for forcing children into extracurricular activities or career paths that do not fit the child's own special blend of talents and interests? Yes and no. With a parent like Roger Camden, gene modifications will accentuate the pressure on children. When parents fail to understand or appreciate the limited role of genes in shaping a child's personality, unrealistic expectations may lead to conflict and coercion. But there is another possibility. In a world where parental expectations are already overbearing, gene interventions may assist parents and children in matching their interests and realizing their dreams. By increasing the likelihood that a child may actually accomplish what the parent wants, gene interventions could reduce conflict between parent and child.

To illustrate this point, let me offer two examples, one hypothetical and one real. Imagine a child, Mary, born into a family with scientific interests and aspirations, perhaps a distinguished line of physicians. If Mary experiences problems in reading and math, both she and her parents face a long and difficult struggle. Although the mother and father were excellent students, Mary is not. She does poorly in school. If her problem is not well understood, she may be subject to years of pressure and blame: "Try harder, Mary. You're not applying yourself." Perhaps her parents will eventually come to understand their daughter's limits, and Mary will also accept the fact that she does not have the abilities her parents expected. In the best of circumstances, she will find a satisfying and creative career in a nonscience field. But just as likely, she will spend the rest of her life in disappointment.

The opportunity for deliberate gene modification could change this scenario. Improved understanding of a child's genetic makeup could increase our skills of diagnosis, helping us to identify learning disorders and problems earlier than we do now. Gene modification could increase the chances that Mary will have the skills that her parents value. If Mary does exhibit interests and aptitudes similar to those of her parents (a key assumption), then, instead of years of struggle and adjustment, she could move directly into the life for which her parents are well equipped to help her. Careful attention to Mary's interests is central to this outcome. In the best of cases, creatively developing her strengths would replace years of effort spent trying to cope with her problems. Psychological studies of gifted children who become successful adults show that these children do not usually achieve high levels of performance without a long and intensive period of training. This training begins early in childhood with warm and loving teachers and parents, who are then supplanted by more demanding master teachers.[9] By identifying and molding Mary's abilities from the outset, her parents could enable her to experience a satisfying life of achievement.

My second example comes from real life and is more complicated. I draw it from the biography by Lance Armstrong's mother, Linda Armstrong Kelly. In the biography, Linda recounts a moment during Lance's teenage years when he was actively involved in triathlon competition, an event combining cycling, swimming, and running. Lance loved the sport, but one day his swimming coach approached Linda and recommended that Lance devote himself entirely to swimming. The coach believed that the young athlete was dispersing his talents, that he needed focus. Linda listened to the coach, and she listened to Lance, who intensely loved the triathlon competition. In the end, she sided with her son. The happy outcome is that Lance, who might have been just a good swimmer, became a champion cyclist.

Linda Armstrong Kelly took from this episode the lesson that parents must pay attention to a child's interests and talents and not just impose their own expectations. "Expecting," she says. "I love that word. That about says it all, doesn't it? From that moment you know you're going to be a parent, you have it all mapped out. And you usually forget to put things like stubbornness and willfulness and an entirely unpredictable sense of style on the chart." She adds, "Maybe the thing about your kid that's driving you crazy isn't a fault at all. Maybe it's just a different aspect of his personality and he needs a little help finding the silver lining to it."[10]

These are wise remarks, and at first sight, they seem to run directly counter to my argument for planning and modifications. Don't impose your expectations; accept your child; above all, pay careful attention to his or her instincts and needs. But setting these remarks in the context of Lance and his mother's relationship gives a slightly different and more nuanced picture. Linda Armstrong Kelly was an intensely driven single mother, always 100 percent there for her child. When Lance became committed to cycling, it was Linda who transported him to distant events, who rooted for him at four in the morning as he raced through remote waypoints, and who dedicated much of her life to developing his cycling career. Her decision to side with Lance, not the swim coach, was based in part on her sensitivity to her son's personality and strengths. For example, she knew that from earliest childhood Lance loved the speed associated with cycling (this became a problem when he took up driving!), and she sensed that swimming would not satisfy him in that respect. She also realized that Lance shared her own stubborn determination and had to be allowed to exercise his will. This sensitivity and deep knowledge of her child, a knowledge partly rooted in her own self-knowledge, made her the boy's best mentor.

How would knowing more about a child's genetically inherited propensities change this picture? Would it increase a parent's understanding of a child's needs, or would it blind the parent to the complex and unexpected novelty of that child's personality? Would it amplify a parent's insight and ability to support, or would it detract from the careful attention to and respect for the particularity of the child that good parenting requires? I honestly do not know the answer to these questions. My point is that they *are* questions, and questions whose answers are by no means clear. Much will depend on whether increased genetic knowledge and control are accompanied by good parental counseling and education. Parents who mistakenly believe that there is "one gene" for "one trait" will greatly overestimate the power of gene modifications to produce results. They will be set up for overexpectation, disappointment, and coercion. But parents who are well counseled and who are led to understand that genes are only one factor in the blend of things that make people what they are may be able to use this knowledge to better understand their child's needs and see that the family meets them.

The full range of answers to these new and challenging questions is years away. If we move in the direction of prenatal gene modifications, we will have to learn what kinds of preparations are likely to maximize their value to children and families and reduce the chances that the interventions become an excuse for coercion. A great deal of research lies ahead. But as we think about these changes, it helps to remember that we already live in a world where parents strenuously try to shape their children's lives. They do so now in a situation of considerable ignorance. For most parents, their children's range of abilities—and vulnerabilities—is a black box, opaque to their understanding. True, parents try their best to pay careful attention to their children. Does my child have athletic interests, or is he musically inclined? Which extracurricular activities does my daughter most enjoy? Parents also draw on knowledge of their own strengths and weaknesses. They suspect that they pass some of these along, whether it is their passion for literature or their aversion to contact sports, and they watch how each child does in these contexts. But their observations and guesses are frequently wrong, and they may look in the wrong direction. Parents can spend years of child rearing missing a child's hidden strengths. The central question, then, is whether the current state of relative ignorance would be improved by the limited insights that genetic analysis and intervention might permit. Whatever the answer to this question, it is sobering to realize that the present way that we bear and rear children also has its problems. We cannot just summon up fears about the future. We also have to weigh the fears of change against a very imperfect present reality.

The insight that our genetic endowment is currently a mystery to each of us is worth keeping in mind as we evaluate the second broad concern about the impacts of gene modification: that deliberate parental interventions could deprive a child, and the adult that child will become, of the deep psychological rewards and satisfactions, the self-esteem, that come from shaping one's own life.

No one has voiced this concern better than the technological critic Bill McKibben. In his book *Enough,* McKibben offers the example of running, which forms an important part of his life. Though not a world-class marathoner, McKibben takes pleasure in pushing his limits and learning how far he can drive himself by means of discipline and effort. When he achieves new levels of performance, he experiences a euphoric "flow state" that he regards as the supreme reward for his exertions. Against this background, McKibben asks how deliberate parental interventions might affect these important psychological dynamics:

> I began by talking about running, because it is one of the contexts I've created for myself, one of the things that orders my life, fills it with metaphor and meaning. If my parents had somehow altered my body so that I could run more quickly, that fact would, as I said, have robbed it of precisely that meaning I draw from it. The point of running, for me, is not to cover ground more quickly: for that, I could use a motorcycle. The point has to do with seeking out my limits, centering my attention: finding out who I am. But that's very difficult if my body has been altered. If the "I" and the Sweat-

works2010 GenePack are entwined in the twists of the double helix. And if my mind has been engineered to make me want to push through the pain of running, or not notice it at all, then the point has truly vanished. My effort to carve out some context for myself is in vain: I might as well be Seinfeld, sitting on his couch and cracking wise about the pointlessness of it all.[11]

This is a strong defense of our current, rather opaque relation to the forces shaping our bodies and minds. But like all expressions of status quo bias, it both misrepresents our current situation and wrongly projects only negative possibilities onto the future. As for the present, McKibben portrays our struggles as a journey of self-discovery in which we triumph through sheer determination. He makes it appear as though biologically influenced capabilities and limits play no role, as though each of us is self-made. In reality, whatever I accomplish is profoundly shaped by my physical and mental inheritance. These parameters, in addition to whatever determination I bring to an activity, already play a major role in shaping outcomes. It is not as though such factors are not at work; it is just that I am imperfectly aware of their presence. As the science journalist Ronald Bailey reminds us, "Human freedom cannot and does not rely on ignorance and randomness."[12] But that is precisely what McKibben seems to believe.

A central question, then, is why a better comprehension of the factors shaping my performance should negatively affect either my effort or my pride in accomplishment. Like McKibben, the President's Council on Bioethics thinks that comprehension has a negative effect. The report *Beyond Therapy* insists that "artificial enhancement can certainly improve a child's abilities and performance . . . *but it does so in a way that separates at least some element of that achievement from the effort of achieving.*"[13] But the council and McKibben greatly overestimate what knowing about our biologically inherited capacities does to our sense of achievement. If I am a good runner in a family of runners, I am already aware that my biological parents have partly given me some of the capacities I draw on: strong bones, taut muscles, a love of exertion. None of this affects my pride in my achievements. How is this changed if they have deliberately given me similar qualities or perhaps some new ones by means of genetic modifications? How is it changed if I know about it? Does it really alter my relationship to my efforts if I have received the "Sweatworks2010 GenePack"? If I know I have been given some special assets for long-distance running, I can feel those strengths at work as I train, just as the natural child of a champion runner does. But I must still bring every resource of my will to develop those gifts. Why does any of this prevent me from achieving McKibben's flow state? How does it deprive me, once I have triumphed in a race, of saying sincerely, "How lucky I was to be able to take my given strengths to new levels of achievement"?

McKibben's error is that he sets up a false dichotomy. On the one hand, he presents the future world of deterministic gene modification as devoid of all human freedom and effort. On the other hand, he depicts a current "natural condition" as one where free will and determination are unaffected by genetically shaped biological attributes. But a world where we know about our genetic endowments does not have to be deterministic. It contains ample opportunity for striving and accomplishment. Indeed, given the knowledge it affords, it may even help channel and focus those efforts. Besides, our current world is not altogether free. As the philosopher Harry Adams reminds us, "Nature unguided does not necessarily leave us any more autonomous with the decisions foisted upon us than does Nature guided."[14] We are already deeply shaped by biological realities. The problem is that we do not understand them well and either take credit for what is not our own doing or fail to see just what we have added to our given talents. Like so many who fear enhanced genetic control, McKibben overvalues ignorance and makes the mistake of confusing it with freedom.

If the previous concern is psychological—whether a child will feel constrained by parental genetic interventions—the third broad concern is ethical: that such interventions somehow violate "a child's right to an open future." Aspects of this right have been developed by bioethicists and philosophers like Hans Jonas, Joel Feinberg, and Dena Davis. Feinberg and Davis both insist that parents are trustees for a child's future exercise of freedom and choice. Parents can intervene to protect the child's future autonomy, but they cannot ethically shape or foreclose that future just to suit their own interests and needs. Davis worries that the choice of a child's genetic traits, whether its sex or other characteristics, wrongly imposes the parent's will on the child.[15] Jonas insists that there is a firm moral principle governing our rearing of children that genetic interventions violate: "Respect the right of each human life to find its own way and be a surprise to itself."[16]

Part of the concern here has to do with the image of parents imposing their own choices or preferences on a child. Isn't this out of keeping with the guardian role, according to which the parent should patiently await the child's development of its mature personality, at most intervening only when strictly needed to preserve the child's future freedom?

The problem with this picture is that it is not an accurate representation of the way most people act as parents and believe it is right to act. As the philosopher William Ruddick points out, parents are both guardians and gardeners.[17] They are guardians because children will grow up to become free and independent persons. This means that, to some extent, parents must protect their children's future interests. But, Ruddick adds, "Parents beget, bear, and raise children to fulfill certain aims, sometimes eccentric or personal. Their productive aims may be reproductive, in the service of continuing some aspect of their own lives through their children."[18] In this respect, parents are like gardeners who raise a crop not just for the plants' sake but for their own sake as well. Ruddick continues: "Children are not their parents' chattels, but they should not be allowed to become their parents' crosses, too heavy for all but self-sacrificial saints to bear. We must find principles that make parenthood a reasonable, rewarding undertaking. Now that parenthood is a matter of choice, rather than a biological or cultural fate for women, we must coordinate parents' desires and children's needs under principles that will guide parental actions with a minimum of regret, resentment, or legal enforcement."[19]

Ruddick observes that as parents and children live together, "there is an interplay and adjustment of need and desire. The child's needs may come to elicit desires to fulfill them." At the same time, "parental desires may elicit new needs in a child that other children do not have, such as the need to be with and to please those particular adults."[20] In response to this two-way dynamic, Ruddick proposes a principle that he believes respects parents' roles as both guardians and gardeners. He calls this the Prospect Provision Principle. It requires parents to foster life prospects (1) which jointly encompass the futures that the parents and those they respect deem likely and (2) which, if realized, would be acceptable to parents and child alike. This principle takes in elements of the parents' wishes *and* the child's future well-being.

It is not hard to find evidence for the soundness of Ruddick's view, as opposed to Feinberg or Davis's claim that we must always act to preserve the child's right to an open future. How else can we explain the myriad decisions which parents make for themselves that often deeply and permanently affect their children's lives? These decisions include the timing of conception, the number and spacing of siblings, the financial and emotional circumstances into which the child arrives, and the educational and religious environments in which the child is brought up. Consider, for example, the offspring of two missionary parents. An article in the *New York Times Magazine* in January 2006 recounts the experience of the Maples, an evangelical Protestant missionary family in Kurungu, northern Kenya. Rick and Carrie Maples and their two daughters, Stephanie and Meghan, moved to this remote mission outpost in September 2004. But Stephanie and Meghan were not sure that northern Kenya was where they wanted to be. Four-year-old Stephanie started to cry right after Meghan said grace one night at dinner. "I miss my friends," she said. Meghan, who was twelve and home-schooled, seemed even less sure than her little sister that she liked being in Kurungu. It wasn't that either girl lacked an intrepid spirit. "Meghan proved her ability to adapt eight years earlier, when the Maples took on their first mission posting, a two-year assignment that they extended into six," reports Daniel Bergner. But it was different in Kurungu, which was much more remote than the first mission. Meghan was struggling with the language and with the lack of girls her own age. "Sometimes I think I can live without friends, I just don't know," she told Bergner. "I didn't really hear God talk to me telling me to be a missionary."[21]

Someday Stephanie and Meghan Maples may thank their parents for their unique Kenyan experiences—or they may not. But whichever the outcome, people generally accept that parents have the right to involve their children in the parents' own most important dreams. Forbidding this, as Ruddick makes clear, would change the very meaning of parenting, deprive parents of some of their deepest aspirations for themselves and their children, and render children crosses "too heavy for all but self-sacrificial saints to bear."

If Ruddick is right, as I think he is, then parents do not violate a child's freedom and autonomy by molding the child's nature in directions shaped by the parents' own hopes for their child, including the use of gene modification. I think that, within limits, parents have the right to impose their dreams on a child. Nor is it clear that genetic modifications must compromise a child's freedom. Joel Feinberg points out a paradox at the heart of child rearing that he terms the "paradox of self-determination."[22] Through socialization and upbringing, says Feinberg, "parents help create some of the interests whose fulfillment will constitute the child's own good. They cannot aim at an independent conception of the child's own good in deciding how to do this, because to some extent, the child's own good (self-fulfillment) depends on which interests the parents decide to create."[23] If I love sailing, for example, and introduce my daughter to it at an early age, offering her the experience and skills and support that she needs to appreciate the sport, I inevitably shape her future interests. Am I imposing my interests on the child, or am I doing what parents always do when introducing a child to the world's possibilities?

Feinberg points out that this more complex understanding of a child's self-determination does not flatly contradict the child's right to an open future. In keeping with this right, he does not believe that parents ought to force their interests on a reluctant child in the course of socializing the child. But he does believe that parents are more than mere guardians of some preestablished set of inclinations in the child. Parents play an active—and unavoidable—role in helping establish, shape, and encourage those very inclinations.

Do genetic interventions fit into this picture? If I enhance my child's aptitudes for certain activities, do I limit the child's freedom or increase it? Let us assume for the moment that some basic interests, like the love of diving, or music, or math, have genetic underpinnings. If so, do I, by undertaking genetic modifications that stimulate these interests and reinforce abilities for them, foreclose my child's right to an open future, or do I merely furnish the child with attributes around which he or she can fashion a life? In this respect, genetic choice does not seem so different from the various forms of socialization that parents undertake. Like parents' choices of educational opportunities, religious upbringing, or recreational pursuits, genetic choices do not run counter to a child's nature so much as provide the foundation on which the child builds a personality.

Some people argue that genetic changes are different from socialization because they seem to be more hard-wired and less easily set aside than the lessons and experiences that parents provide. A child would seem to have less freedom to reject these parental impositions and forge a unique plan of life. Is this true? Some forms of early socialization are extremely powerful. The oft-quoted Jesuit remark "Give me the child until he is seven and I will show you the man" illustrates this notion.[24] It also is not clear that many gene interventions would intrude in any way on a child's freedom. If I give my child enhanced capabilities for music, or math, or sports, the child remains free to apply (or not apply) these talents in any way he or she wishes. I may dream that my son or daughter will become an astronaut, but the child may choose to apply the given analytical skills to a business career. In other words, most gene modifications will be enhancements of basic, all-purpose capabilities that augment a child's abilities to move in any direction. Even if I were able to use genetics to strongly bend a child's interests in my wished-for direction, it is not clear that this forecloses the child's freedom. If the child accepts those interests, his or her freedom is unimpaired. Only if the child experiences conflict from other, competing tendencies do my gene modifications create a problem. In that case, the child will have to make difficult choices to find the best course. But even in this case, the child will be like most of us, trying to turn an array of inherited and sometimes conflicting desires and abilities toward creating a satisfying life.

Once again, it is important to recognize that genetic tendencies already shape who we are. It is not as though we are inscribing parentally introduced genetic influences onto a blank slate that the child would otherwise be free to write on as he or she wished. A child will have biologically influenced talents and interests no matter what parents do, and some of these will exert a potent influence on the child's life. Unless we consider the random lottery of nature's bestowal as somehow less restrictive than parents are, it is hard to see how a parent's choice of some of these attributes limits a child's freedom. It would do so only if parents were to follow up their genetic choices by ignoring the child's actual array of qualities and interests and insist on imposing their preferences regardless of their child's wishes. But this is bad parenting in any case and does not necessarily represent our future experience with well-guided gene interventions. As the bioethicist Glenn McGee reminds us, "The key is to avoid extreme measures through biological *or any other means,* and to temper decisions before birth with the recognition that every child has a right to make some decisions about her own identity."[25]

The fourth broad concern is that parents' selection of a child's genetic attributes could cause a moral distortion of the whole way that we relate to other human beings, leading to a replacement of respect for persons with a valuation of people based on superficial traits. This is a concern already articulated by disability theorists. Adrienne Asch, a spokesperson for disability rights, criticizes any type of trait selection, whether to eliminate a disability or to enhance traits, because she sees it as committing the fundamental moral error of making one trait or quality stand for the whole person. In Asch's view, to select or discard an embryo or terminate a pregnancy just because it possesses (or lacks) some trait is to reduce the whole complexity of a human life to a single feature. "Ending an otherwise desired pregnancy after learning of a diagnosis of spina bifida or cystic fibrosis says that this one fact trumps everything else one could discover about the child-to-be, and says that the woman (or couple) cannot accept into her intimate life a child with this characteristic when she planned to accept a child. A health report card becomes precursor to membership in the family, making the family rather like 'the club.'"[26]

Because Asch believes that the family is a school for our moral education, she also thinks that trait selection and exclusiveness will ultimately undermine society's willingness to accept all people and treat them with respect.

This concern has a grain of truth. If I relate to other people solely in terms of whether they possess some quality—IQ, attractiveness, sightedness, mobility—I show myself to be a superficial and even morally repugnant individual. If parents evaluated and loved their children in terms of how well they measured up to some preset menu of characteristics or accomplishments, family life would be unbearable.

But before we know someone well, when we are creating a relationship, it is usual to search for certain desired traits. For example, in seeking a mate, I might go to a cultural event—perhaps a book signing—where I hope to meet someone who enjoys reading as much as I do. Since I am not keen on winter sports, I probably would not attend a singles weekend at a ski resort. Would anyone declare me to be morally superficial because I tried to find a book-loving partner rather than a skier? Would they say, "You are so shallow. You shouldn't be picky. Try to accept people as they are"?

The philosopher Frances Kamm has presented a helpful distinction between "caring about" and "caring to have."[27] In matters of the heart—and this includes romantic attraction, marriage, and parental love—most people rightly feel that they should *care about* someone whether or not that person has certain traits. We regard it as morally disturbing when, after years of marriage, a husband rejects his wife because she is aging. But we do not regard it as odd to choose a date based on youthful vitality, hair color, or other rather superficial qualities. Nor is it odd for people struggling with infertility to choose a sperm or egg donor because of that individual's surface qualities. In

Kamm's terms, we should always *care about* the people with whom we have entered into significant human relationships. But in creating those relationships nothing prevents us from *caring* for them *to have* certain qualities we appreciate.

Another *New York Times Magazine* story illustrates this point. The story, entitled "Wanted: A Few Good Sperm," explores the rapidly growing world of single women who are choosing to have children with the help of anonymous sperm donors.[28] Here is how the author, Jennifer Egan, describes one woman's search for a sperm donor:

> With online dating, friends used to say: "What about him? What about him?" I'd say: "Don't like the nose. Ah, the eyes are a little buggy. He really likes to golf, and you know I don't like golfing."
>   . . . Because she herself is so tall, she preferred a medium height. . . . She was also attracted by the idea of a donor of another race. "I believe in multiculturalism," she said. "I would probably choose somebody with a darker skin color so I don't have to slather sunblock on my kid all the time. I want it to be a healthy mix."

This woman's search is amusing, but it does not seem morally offensive. Like most parents of children produced with the help of sperm or egg donation, she will probably love her child no matter how it turns out. As Kamm reminds us, there is a world of difference between caring to have a child with some trait and caring about that child once the child is born. But only the former applies at the level of trait determination. In the future, we may grow used to parents more actively trying to have a child with certain desired traits—and intensely loving whatever child results.

I began this chapter with the question of how gene modification, enhancement, and trait selection might affect parenting. Although I have not exhausted every dimension of the question, I have addressed some of the leading concerns. I have argued that we do not have to worry that trait selection will replace parents' unconditional love with critical scrutiny, because the very psychological principle that we fear damaging is also the best protection against its happening. Parental love will almost always prevail. Against the fear that gene selection will turn parents into oppressive taskmasters, I think that it is equally possible that gene selection may improve the resources that parents have at their disposal for understanding and guiding their children, and may reduce some of the unreasonable pressures that parents now exert.

I believe that we do not have to fear that genetically modified youngsters will lose their sense of self or justifiable pride in their achievements. Bill McKibben fails to see that there is a role for personal accomplishment in everything we do with our inherited biological characteristics. He also confuses ignorance about our inherited traits with a state of freedom. Although ignorance may offer the comforting illusion of self-determination, it is by no means clear that it is a better basis on which to build our lives than one allowing greater insight into our limits and strengths.

The claim that children have a right to an open future that is violated by gene selection does not stand up when measured against accepted parental practices. Although parents should respect their children's future freedom, they also have the right to shape their children in directions influenced by the parents' own dreams. Parents are both guardians and gardeners. Shaping a child's abilities, inclinations, and talents can be consistent with respecting a child's freedom, since these inclinations and talents provide the foundation for the use of freedom. Finally, I have tried to show that trait selection is not incompatible with the full respect we owe other human beings and those we love. We can care about people even if, in establishing relationships with them, we care for them to have qualities that we desire.

Because the family is the center and source of so much that we value, we should be worried about anything that threatens it. But many of our fears are overblown. Consider this. Less than a century ago, in my grandmother's day, almost every decision about becoming parents, including the timing, spacing, and number of children, was left to chance. Today we use contraception to determine how many children we will have and when we will have them. If we face infertility, we use high-tech medicine and test-tube conception to procreate.[29] People needing the help of sperm or egg donors choose those donors from catalogues providing detailed information about the donors' traits and background. Women routinely undergo dozens of prenatal tests to ensure the health of the child. Following birth, we do further tests on the child and closely monitor the child's development to catch and correct developmental problems.

Has any of this impaired parenting? Do we reject our children more often than people in my grandmother's generation did? Do we impose unreasonable expectations on them? I doubt it. It would be hard to prove that the quality of parenting today is worse than in the past. I expect that as we move into the era of human gene modification, parental love will continue to flourish. The family is a resilient institution and is still our best protection against the things we fear might damage it.

CHAPTER 6

Will We Create a "Genobility"?

In *Remaking Eden,* the Princeton University biologist Lee M. Silver offers a troubling prediction about a future world where gene enhancement has torn the human race apart. Sometime in the distant future, says Silver, "humans would diverge into just two species, the GenRich and the Naturals. Naturals had the standard set of 46 chromosomes that long defined the human species, while the GenRich alive at that time had an extra pair specially designed to receive additional gene-packs at each new generation. With 48 chromosomes and thousands of additional genes, the GenRich were, indeed, on their way to diverging apart from the Naturals."[1]

Silver's prophecy is not new. Over a century ago, the great British science fiction writer H. G. Wells, in his 1895 novel *The Time Machine,* followed a time traveler to a remote future where the human race has become separated into two species. The Eloi are a gentle, herbivorous people who live on the parklike surface of the planet. Dwelling beneath them, in a dark subterranean world, are the Morlocks, a grotesque, mole-like species, who maintain the planet's mechanical life support systems and who, as their price for this service, occasionally capture and devour an Eloi.

Wells's two species are not the result of genetic engineering. They seem to have evolved naturally over time from the extreme class divisions in nineteenth-century British society. Wells's warning to his contemporaries is clear: Don't allow economic injustices to become so acute that they ultimately transform human biology. Today his warning takes on new meaning. It tells us not only to resist gross economic injustices but also not to allow them to become connected with the use of new biotechnologies in ways that reinforce the gap between society's haves and have-nots. Above all, it tells us not to let social differences become allied with biology in ways that culminate in speciation: the separation of humankind into different breeding populations.[2]

The negative impact of genetic manipulations on society is one of the leading themes of critics of gene manipulation, who raise at least three different concerns. First, they worry about what will happen if the affluent use gene technology to gain an advantage over the poor, both within nations and on a global basis. Will this create a growing and irreparable divide between social classes and between poor and rich nations? Will it, in the most extreme case, as depicted by Silver, lead to speciation? Second, they worry that gene enhancements could erode our basic respect for one another. Will the ability to confer socially powerful genetic enhancements on one's children harden the attitudes of social classes toward one another? Will it lead me *and* my children to see ourselves as fundamentally different from the less fortunate? Will we come to regard ourselves as entitled to the social and economic privileges we enjoy rather than, as now, accept that much that we have results from the random throw of nature's dice? The critics find evidence for both these fears in the terrible experience of the eugenics movement, when genetics was used to buttress social oppression and exclusion.

Finally, some critics worry about the immediate impact on the poor. They ask how we can possibly justify spending millions of dollars of private or federal research funds on high-tech genetics in a world where so many people lack access to basic medical care. Inadequate prenatal care already jeopardizes the lives and health of millions of children. How can we think of devoting scarce resources to gene enhancements for a privileged few?

These are powerful concerns and objections. They highlight some of the worst things that could happen if access to gene modifications were left to market forces. But like other fears, they develop only one side of the story. Gene enhancements, if properly handled, could narrow the gap between society's haves and have-nots and between developed and developing nations. Although gene modification could lead to new forms of pride, it could undermine some of the arrogance that now buttresses class and social divisions, and it could create new opportunities for many at the bottom of the social ladder. It could increase rather than reduce parents' reproductive options, and it could avoid the eugenics abuses of the past. And even though expenditures on gene research should not replace the provision of basic health care, it may be wrong to see funding as a zero-sum game, where money put into gene research comes out of the pocket of basic medical care. Like some other advanced technologies, germline gene modifications may actually *lower* health care costs for everyone by replacing costly halfway medical technologies with dramatic new solutions to health care needs. As is true of some other concerns explored here, the lesson is not that the fears are unjustified but that they should not replace careful thinking about whether they can be managed and overcome.

Fears about the impact of gene enhancements on social divisions are not hard to understand. Because the earliest efforts at inheritable gene interventions will be costly, they are not likely to be funded by the government. Both the research and the initial clinical uses are likely be financed out of private resources—for example, infertility clinics seeking to expand their repertoire of services. They will be made available to those who can afford them. People of modest means, like Antonio and Marie in the film *Gattaca,* will be shut out, and only the affluent are likely to receive the benefits for their children. But since many affluent people already possess some skills and talents that have helped propel them forward economically, their children's gains will build on gains. Reduced exposure to disease and advanced reading or math skills in one generation will be enhanced in the next. Eventually, economically valuable talents, like athletic, verbal, or artistic abilities, will be passed on and improved. It is true that gifted parents now transmit some of their abilities to their children. But ordinary sexual reproduction is a gamble. The most talented parents often have average children, and ordinary people can produce geniuses. Over a century ago, the pioneering geneticist Francis Galton introduced the phrases "regression towards mediocrity" and "regression to the mean" to describe how nature's reproductive lottery tends to level extreme abilities across time. Deliberate gene enhancements provide a way for affluent parents to halt or even reverse this regression. All these possibilities are intensified if a society already displays acute economic and social differences.[3]

No one has described a likely scenario of the negative impacts of gene enhancements on society better than the lawyer-bioethicist Maxwell Mehlman, who helped introduce the term "genobility." Mehlman is not unalterably opposed to gene enhancement. But in his 2003 book, *Wondergenes,* he describes a possibly fatal slide that could start with an early generation's wealth-based access to genetic enhancements.[4] At the start, says Mehlman, these enhancements could benefit everyone through the increase in the number of talented people. But in time, a burgeoning underclass may develop that initially cedes power to its genetic superiors "in return for the material benefits made possible by genetic advances." Political democracy may persist for a while, with the unenhanced electing representatives of the genetic upper class and the new genetic-political-economic elite ruling "according to enlightened principles of *noblesse oblige*." The resulting state is likely to be highly unstable, however. A quasi-democratic system would be highly vulnerable to demagogues who promised to redistribute social benefits or genetic enhancements more evenly. The enhanced elite, in turn, would be reluctant to accept inroads on their privileges and power. In the end, society would be likely to swing in "ever-widening arcs between rule by underclass demagogues and rule by a genetic aristocracy," with possible deterioration into mob rule and anarchy and with the unenhanced going so far as "to destroy the scientific foundations of the genetic revolution." Or, what Mehlman calls "post-geno-revolutionary society" could devolve into totalitarian rule by the genetic elite, who use repressive techniques to retain control. Mehlman concludes with the observation that the "ultimate horror of genetic enhancement is not that we will create monsters. It is that we will create gods."[5]

Francis Fukuyama worries about the changes in psychology that might come over genetic elites. Fukuyama maintains that the political rights of citizens in the Western tradition have rested on the belief that, regardless of our differences, human beings share a common nature "that allows every human being to potentially communicate with and enter into a moral relationship with every other human being on the planet." For Thomas Jefferson and the other founders of modern democracy, the political equality enshrined in the Declaration of Independence rested on the empirical fact of natural human equality. Fukuyama quotes Jefferson's summation: "The mass of mankind has not been born with saddles on their backs, nor a favored few booted and spurred, ready to ride them legitimately, by the grace of God."[6] But what happens if growing genetic differences split the human race into one or more very different biological subspecies? Will we be able to sustain our commitment to universal human rights? "The ultimate question raised by biotechnology," says Fukuyama, is, "What will happen to political rights once we are able to, in effect, breed some people with saddles on their backs, and others with boots and spurs?"[7]

Fukuyama's worries are perplexing. American and British democracies both evolved in cultures where there was widespread belief that human beings differ greatly in their natural abilities. The Declaration of Independence may assert it to be self-evident that all men are created equal, but that claim was largely a moral rather than a natural one, and it was often made in the face of widespread belief in a natural hierarchy of talent. As Ronald Bailey observes, "The modern ideals of democracy and political equality are sustained chiefly by the insight, developed by Enlightenment thinkers, that people are responsible moral agents who can distinguish right from wrong, and therefore deserve equal consideration before the law and a respected place in our political community. The broad ability to distinguish right from wrong does not depend on the genetics of IQ, skin color, or gender. With respect to political equality, genetic differences are already differences that make no difference."[8] In view of this compatibility between moral equality and genetic difference, Fukuyama's worries seem unfounded.

The philosopher Michael Sandel, like Fukuyama a member of the President's Council on Bioethics, goes much deeper than Fukuyama. He sees genetic enhancements as potentially imperiling some of the vital interior sentiments that sustain the democratic commitment to social justice. Sandel develops his concerns in an influential article that appeared in the *Atlantic Monthly* in 2004 entitled "The Case against Perfection."[9] He observes that our commitment to social justice rests on "three key features of our moral landscape: humility, responsibility, and solidarity." We have these attitudes, Sandel says, because we know we have to share our fate with others. Take the matter of medical insurance. Because we do not know how or when various ills will befall us, we pool our risks by buying health insurance and life insurance. Even without a sense of mutual obligation, our common vulnerability leads us to share one another's fate.

But what happens if my children and I can permanently escape much of the shared human condition by gene improvements that leave our less fortunate neighbors behind? Because these accomplishments occur through the choices my family has made, we can easily see ourselves as somehow "self-created." Is my family privileged and comfortable? "Yes," we might say. How did we get here? "By our hard work over many generations, and by the astute use of the new genetic technologies?" What about the less fortunate? "Too bad. They had their chance. Their fate is their own doing." Sandel states the problem as he sees it here:

> A lively sense of the contingency of our gifts—a consciousness that none of us is wholly responsible for his or her success—saves a meritocratic society from sliding into the smug assumption that the rich are rich because they are more deserving than the poor. Without this, the successful would become even more likely than they are now to view themselves as self-made and self-sufficient, and hence wholly responsible for their success. Those at the bottom of society would be viewed not as disadvantaged, and thus worthy of a measure of compensation, but as simply unfit, and thus worthy of eugenic repair. The meritocracy, less chastened by chance, would become harder, less forgiving. As perfect genetic knowledge would end the simulacrum of solidarity in insurance markets, so perfect genetic control would erode the actual solidarity that arises when men and women reflect on the contingency of their talents and fortunes.[10]

Like so many expressions of concern about gene enhancements, both Fukuyama's and Sandel's assessments start from an uncritical acceptance of the status quo. Both writers presume that our current state of genetic vulnerability supports respect for others, and they see these sentiments as threatened by increased control. Small wonder that both writers have been leading voices on the President's Council of Bioethics, where their arguments have influenced council opposition to gene enhancements. Very shortly, I shall ask whether the opposite of these attitudes may be true: whether our current absence of genetic control sustains sentiments of pride and moral indifference. But before I get to that, I want to present a scenario that to some extent supports Fukuyama's and Sandel's views, one developed by Nancy Kress in her *Beggars* series of novels.

In chapter 5 we met Kress's main protagonist, Leisha Camden, one of the first generation of Sleepless children. I described the harsh attitudes of Leisha's father and his rejection of Leisha's natural sister, Alice, but I said nothing about his political philosophy. In fact, the portrait that Kress offers of Roger Camden provides imaginative support for Sandel's argument. Roger is a disciple of Kenzo Yagai, a brilliant Japanese scientist who has invented a method for cold nuclear fusion known as Y-energy that brings the world unlimited power and prosperity. Yagai and Roger advocate an Ayn Rand–like doctrine of reward for individual merit and posit the wrongfulness of reliance on the "generosity" of others. They believe only in the free, uncoerced trade of one's assets. Those who have no assets, and hence nothing to trade, must accept their fate. Unfortunately, such people, the "beggars" of the trilogy's titles, too often resort to theft and violence.

In the first volume of the trilogy, while Leisha is still a child, she has a fateful conversation with her father. He tells her, "You're special. Better than other people. Before you were born, I had some doctors help make you that way."

"Why?" Leisha asks.

"So you could do anything you want to and make manifest your own individuality. . . . Excellence is what counts, Leisha. Excellence supported by individual effort. And that's *all* that counts."

Leisha replies, "I'll make my specialness find a way to make Alice special, too."

To Roger, Leisha's compassion for her unmodified sister, Alice, misses the point. People do not help others. Only contracts and the free trade of assets make sense. In Roger's Yagaiist / Ayn Rand–like philosophy, there is no such thing as kindness, compassion, or human solidarity. Roger's commitment to gene enhancements used for personal benefit both reflects and reinforces his underlying philosophy of individual achievement and entitlement.

The divide between Roger's philosophy and Leisha's generous instincts shapes the narrative in the first volume of the trilogy. Leisha proves to be a brilliant student, eventually attending Harvard Law School. While she is still a young girl, scientists discover that there is a further benefit to sleeplessness. Although it evolved to protect humans against nocturnal predators, sleep is toxic. Because the Sleepless are free of its effects, they do not age. Leisha, in her sixties, following a successful legal career fighting discrimination against the Sleepless, establishes a small community in New Mexico aimed at educating talented Sleeper children. She remains vigorous into great old age, when she falls victim to the growing social turmoil. Following an air crash, Leisha is slain by a group of fundamentalist Sleepers opposed to all gene modifications.

Other Sleepless take an opposite course from Leisha's. Jennifer Fatima Sharifi, a Sleepless heiress to a huge fortune, becomes increasingly concerned about what she sees as the rising tide of Sleeper envy, greed, and discrimination. She purchases a huge tract of land in Pennsylvania, which she develops as a fortified "Sanctuary" for the Sleepless, who are asked, as the price of admission, to swear absolute allegiance to the community. In the mid-twenty-first century, as tensions continue to grow between the two communities, Jennifer seeks to escape the Sleepers' hold by moving Sanctuary to an orbital city-state in space.

In the new haven, Jennifer and her followers take genetic engineering one step further to produce a new generation of "Super" offspring with extreme "genemod" intelligence enhancements. This generation is intended to be the way around genetic regression to the mean, ensuring that the Sleepless continue to expand their superiority over Sleepers. She has two Super grandchildren, Tony and Miri (Miranda). Though brilliant, they stutter and display twitching motions, which makes them seem odd to many of the normal Sleepless ("Norms"). Miri grows up feeling unloved and ugly. When Tony is injured in a fall from Sanctuary's center, Jennifer Sharifi orders him to be euthanized, on the grounds that there is no room for dependency in a Sleepless world. (She had previously done the same thing with a congenitally damaged infant.) This deed outrages Miri and the other Supers, who see it as a betrayal of the nonviolent elements of Yagaiist philosophy.

Jennifer moves toward the culmination of her career: a declaration of independence from the beggars of Earth. To accomplish this, her scientists have deposited a short-lived but lethal genetically modified virus in terrestrial cities that can be actuated on her command. A military confrontation looms with the U.S. government, which continues to see Sanctuary as within its national borders.

At the last moment, the small group of Super children intervenes. Their superior intelligence allows them to take control of the orbiter's systems, and they suppress the rebellion. The SuperSleepless, who were brought into being to perpetuate Sleepless power, instead betray their parents, sending them to prison for treason. The first novel in the trilogy ends as Miri and some other Supers join Leisha's small community in New Mexico, where they aspire to a more communalistic and artistic-intuitive style of life while respecting the ever-changing differences among the forms of human beings.

Eventually released from prison, Jennifer Sharifi, the epitome of Sleepless arrogance and domination, continues her malevolent role in the other two novels in the series, *Beggars and Choosers* and *Beggars Ride*. As the Sleepless withdraw their technical or financial support, Sleeper society spirals downward into a state of near paralysis. In collaboration with a renegade Russian gene scientist, Serge Mikhailovich Strukov, Jennifer disseminates a new virus designed to control the Sleeper population by rendering them fearful and averse to novelty and solitude. She intends to use Strukov's virus to forever eliminate Sleepers' threat to Sanctuary's dominance and to "freeze the world into biological inhibition." But new heroes arise to combat her. They include Lizzie Francy, a "natural," unmodified child of poor Sleeper parents who develops and uses her talents as a computer hacker to successfully defend her imperiled community. Lizzie is joined in her struggle by Theresa Aranow, a gene-modified daughter of wealth (and admirer of Leisha Camden), who is impressed by the devotion and solidarity of Lizzie's Sleeper family and community.

The trilogy ends when Jennifer kills her own granddaughter, Miri, in a nuclear attack, but is herself annihilated in her orbital sanctuary by a nuclear weapon launched by her power-hungry ally, Serge Strukov. Just before dispatching the missile, Serge sends a message to Jennifer quoting the French writer La Rochefoucauld: "The real way to be fooled . . . is to believe oneself better than others."

Much of the narrative of the *Beggars* trilogy illustrates the fears expressed by Mehlman, Fukuyama, and Sandel. Here is society in the chaos that Mehlman depicts, with a privileged governing elite wavering between periodic abandonment of a dependent, unmodified population and tyrannical—not to say genocidal—rule over them. In the increasing disenchantment and hatred of the Sleepless for the Sleeper majority, we see the realization of Fukuyama's fear that a divergence in human biology could lead people to abandon their sense of shared human rights. And in Roger Camden and Jennifer Sharifi's Yagaiist philosophy, we see the full embodiment of Sandel's concern that inheritable gene modifications will erode human solidarity and strengthen the conviction of the privileged that they are self-made and self-created.

Yet in some ways Kress's trilogy also subverts the arguments of the critics. Not all the Sleepless give themselves over to pride and separateness. From childhood, Leisha Camden reaches out to the less fortunate and fights to preserve the unity of the human community. She is joined by people like Theresa Aranow, who resist the slide into violence that both technology and Yagaiist philosophy have produced. Nor are the Sleepers entirely deprived of gifts. In Lizzie Francy we see a genetically unmodified individual who has the computer skills and humanity needed to marshal her community and successfully resist domination by Jennifer Sharifi and her followers. Finally, the separationist and domineering impulses of Sharifi ultimately prove self-destructive. A philosophy that resists community and prizes power undoes itself by inviting similar treatment at the hands of others who profess it.

The *Beggars* trilogy reflects its author's ambivalence about the worth or viability of the meritocratic attitudes that, in Roger Camden and others, initiate the cycle of gene enhancements. Speaking of Ayn Rand's philosophy in an interview, Kress remarks that "although there's something very appealing about her emphasis on individual responsibility," her philosophy "lacks all compassion, and even more fundamental, it lacks recognition of the fact that we are a social species and that our society does not consist of a group of people only striving for their own ends."[11] In the trilogy, deep social impulses and unavoidable interdependence act to combat efforts toward separation. Not only do the gene enhanced continue to have relationships with their unenhanced inferiors, efforts to separate the two communities ultimately prove unworkable and destructive. In the end, what the philosopher Immanuel Kant once called our "unsocial sociability," the unavoidable mutual need and mutual dependence that drive people together, helps to heal some of the wounds that people like Jennifer Sharifi inflicted.

The *Beggars* trilogy is fiction. Its modestly optimistic conclusions may not accurately describe what will happen once people like Roger Camden and Jennifer Sharifi start choosing gene enhancements for their offspring. The far more apocalyptic visions sketched by Mehlman and Wells may be better descriptions of our gene-enhanced future. But fiction has its uses. What the *Beggars* trilogy tells us is that there are forces at work in human thinking and society that resist division and pride. These forces may help prevent disaster. Or, in a best-case scenario not fully sketched by Kress or other writers, they might even convert the powers produced by genetic technology into new and positive opportunities for the human community.

How might a more optimistic scenario develop? In what ways can the new genetic technology work to *reduce* social divisions? My answer starts with a classroom experience that I had more than thirty years ago. In my first year as an ethics graduate student at Harvard, I had the good fortune to enroll in a course taught by a new philosophy professor named John Rawls. I do not think that anyone in the class realized that by the time of his death nearly thirty years later, Rawls would be regarded as the most distinguished moral and political philosopher of the twentieth century. His 1971 book, *A Theory of Justice*, is a classic in the field.[12]

As I listened to Rawls's lectures, I was particularly impressed by his core moral vision. Ethics, Rawls said, represents the effort to bring reason to bear on human social life. Its goal is a society whose members can freely agree on the basic principles that should govern access to power, authority, goods, and privileges. Without such agreement, only conflict or force reign. But how can we attain a society based on free agreement? Rawls's answer involves the idea that we should think of ourselves as entering an imaginary conference room where we can propose, deliberate about, and finally vote on the governing principles for our society. As an expression of the commitment to free consent by all, every participant has one vote. Rawls calls this setting the "original position."

We are permitted to try to do as well for ourselves as we can, but on one condition: we must put aside all the special features that let us know we are different from one another. We do not have to forget that we are human beings: that we have a human biology, that we are physically vulnerable, that we have a sex (male or female), that we are born and die. But we must put aside all the special features that distinguish us from one another. In my case, this means that I must somehow forget that I am a white male tenured professor, and so on.

Rawls terms this perspective of radical impartiality the "veil of ignorance." He believed that adopting this veil was a necessary *conceptual* step for us to take in trying to identify the basic principles of social justice. Assuming the veil of ignorance frees us from fixation on our immediate advantages and disadvantages, and it allows us to engage in long-term thinking about what is good for everyone. Unless the fortunate adopted this perspective, they would try to impose their conditions for agreement on the less fortunate. The less fortunate would resist any resulting principles, and society would always be on the verge of anarchy and violence. Rawls believed that the "original position" and the "veil of ignorance" were ideas that expressed the "fair terms of agreement" for a society that could elicit the free consent of all of its members.

In *A Theory of Justice,* Rawls applies this perspective to developing the basic principles for a just society. He identifies two such principles. The first, the "principle of equal liberty," guarantees every person "an equal right to the most extensive liberties compatible with similar liberties for all." The thinking behind this principle is evident: since the original position deprives me of the knowledge of my special strengths and weaknesses, it makes sense for me to protect my basic civil liberties, such as the right to vote, the right to exercise freedom of speech and conscience, and the right to due process in legal proceedings.

Rawls's second principle applies to society's economic resources. He calls it the "difference principle." It permits inequalities in ownership of material goods and the exercise of authority: different salary levels, different amounts of property, differences in power and authority in private and public institutions. But it says that all such inequalities are permitted only if they are to the greatest benefit of the *least* well-off members of society and if everyone has an equal opportunity to become one of the most well-off.

Given the original position of equality as a starting point, Rawls's theory is broadly egalitarian: equally placed people are likely to seek equal access to life's most vital goods. Why, then, does Rawls believe we should permit the inequalities allowed by the difference principle? Part of the answer is that life is not equal. People vary greatly in their talents—their skills, gifts, creativity, energy, and drive. Some of this is genetic and some familial; some comes from the larger environment; some is the result of the decisions and choices that people make; and some is mere luck. As a result, we have widely different levels of ability to acquire goods and power. Even if we are placed at the same starting point in life's race, some of us will be winners and some losers.

Faced with this reality, people in the original position have two choices. One is to let life's race be run and to accept the outcome. If someone has been given great gifts in nature's genetic lottery, high intelligence, say, or strong verbal abilities, so be it. That person is also likely to be a winner in life's race. If someone else is born with problems—for example, very short stature, a low IQ, or a learning disorder—too bad. That person will be a loser twice over: first, by being deprived of socially esteemed qualities, and second, by falling behind in the economic race. But allowing this to happen goes against everyone's basic interest in the original position of doing as well as they can for themselves and their children. Rawls also observes that there seems to be something unfair about letting luck or nature determine the most basic wins and losses in life. No one in the original position of equality wants the sheer contingencies of life to shape his or her entire future.

As a second option, everyone could be Robin Hoods, taking (almost) everything from the rich and giving it to the poor. This would be a radically redistributive approach. Let people run the race and amass their winnings, but, as they do, take everything extra away from them and give it to the less fortunate. From the standpoint of the original position, this initially makes a lot of sense. Among other things, it compensates me if I happen to have been dealt a bad hand by my genes or circumstances. But ultimately, says Rawls, this strategy is self-defeating. People are often unwilling to contribute to the social pot unless they are rewarded. Human beings tend not to perform at their best without incentives in the form of material or other rewards. Furthermore, when it comes to power and authority in human organizations, not everybody can hold an equal position. Some people have a talent for leadership and some do not. If people are deprived of all the rewards for their efforts or the power they need to achieve things, they will not exert themselves, and *everyone* will be worse off. Anything that undermines incentives can end up destroying the very productivity whose benefits everyone would like to have part of. So this radically egalitarian redistributive approach is no better than one that just lets life's race be run.

The difference principle is Rawls's alternative to these two unacceptable options. It permits differences in rewards and power, but only on the condition that the differences work out to the advantage of everyone in society, especially the least fortunate. Rawls's theory— and the equal standpoint of the original position—makes it wise to be especially attentive to the position of those who have been deprived by birth, family circumstances, or life's course of many of the things needed for achievement. In choosing social policies, we want to protect ourselves as much as we can should we happen to suffer misfortune. But this does not mean we that we should ruthlessly pillage the assets of the more gifted. For one thing, we could also be these people. For another, anything that undermines incentives may be self-defeating. What we want to do is let gifted people exercise their talents and receive the rewards of doing so, but only so long as each increment of reward produces new resources and opportunities that benefit everyone, especially the least well-off. What we do not want to do is remove their incentives to produce. Neither do we want to permit the well-off limitless rewards that have no connection with the fortunes of the least well-off class of people, for we might end up one of the least well-off. So, if paying the CEO of a large company ten times the average wage of one of his workers creates new products or better employment opportunities for everyone, or generates tax revenues that help the less fortunate through improved schooling, medical care, and the like, the difference principle permits this. But it does not permit wages twenty or one hundred times greater than an average worker's if the extra money has no added beneficial effect on the less fortunate. If market forces award such wages, then a just society should have tax policies that transfer that unnecessary increment back to those on the lower rungs of society's ladder.

Rawls's theory has powerful implications for our thinking about social justice, but in at least one respect, advances in genetics have made the theory out of date. Facing the issue of how to handle the difference we experience as a result of birth and our family environment, including genetic differences, Rawls's difference principle tells us to accept them and try to utilize them for social benefit. If by birth or upbringing a Bill Gates has keen skills as a computer scientist, let's permit him to develop those talents and enjoy the rewards of doing so. But let's also make sure that his winnings benefit people who are much less fortunate. (I will leave it to others to decide whether Gates's vast wealth combined with his charitable giving to programs aimed at promoting world health meets the test of Rawls's difference principle.) Rawls offered this approach in 1971 because it seemed obvious to him that we could not easily redistribute many of the inherited talents and skills that underlie social achievements. The best we can do is redistribute some of the material consequences of those abilities.

To a limited extent, Rawls tried to imagine a role for genetic interventions. In a passage in *A Theory of Justice,* he suggests that the genes themselves may be subject to redistribution:

> I have assumed so far that the distribution of natural assets is a fact of nature and that no attempt is made to change it, or even take it into account.... I shall not consider questions of eugenics, confining myself throughout to the traditional concerns of social justice. We should note, though, that it is not to the advantage of the less fortunate to propose policies which reduce the talents of others. Instead, by accepting the difference principle, they view the greater abilities as a social asset to be used for the common advantage. But it is also in the interest of each to have greater natural assets.... In the original position, then, the parties want to insure for their descendants the best genetic endowment (assuming their own to be fixed). The pursuit of reasonable policies in this regard is something that earlier generations owe to later ones, this being a question that arises between generations. Thus over time a society is to take steps at least to preserve the general level of natural abilities and to prevent the diffusion of serious defects.[13]

This fascinating passage shows that Rawls saw genetic interventions as worth considering within the context of a theory of justice.

It also shows a willingness on the part of at least one leading theorist to rehabilitate the concept of eugenics, something I shall return to shortly. Finally, it shows that Rawls himself was somewhat unclear about what genetic science brings to our concern for a more just society. Is the goal merely "to prevent the diffusion of serious defects"? If so, genetic interventions would be confined to the kind of prenatal testing and genetic diagnosis now prevalent, as well as the introduction of germline or somatic gene therapies aimed at reducing the incidence of genetic disorders. Or, is the goal to ensure our descendants "the best genetic endowment"? If so, then we have entered the realm of gene enhancements as an instrument of social policy and social justice. The fact that Rawls is somewhat fuzzy here is understandable. Just a few decades ago, no one could have imagined the progress that genomic science has made in helping us understand the genetic bases of human abilities. Today we can begin to envision a future where underlying abilities might be improved. Taking both of Rawls's suggestions seriously, a just society may want to correct genetic conditions that impair people's social performance *and* provide genetic enhancements that promote success.[14]

Let's consider what this could mean in terms of reading ability, which has become vital to success in modern society. In the past, when society was more agrarian, literacy was less important. Someone who was a poor reader could still be a successful farmer, craftworker, or homemaker. This may be one reason why so many people experience reading problems: skills that are now crucial were once less so, and society has changed much more rapidly than our biology. Because of these social changes, children with reading difficulties now face an uphill battle. For many of them, early failure in school is followed by dropping out and by drifting into self-destructive forms of behavior, ranging from violence to unwanted pregnancy. Failure generates failure. True, a just society tries to combat this. Programs like Head Start aim at catching problems early and correcting them. But such efforts are limited by a child's underlying abilities. As the uncle of a nephew who struggled throughout his life with dyslexia, I can personally testify to how much this disorder can handicap even the brightest and most motivated young person.

The gene variations that contribute to dyslexia and other reading problems are now being identified, and more will be in the future. Imagine that we could identify this condition in a family's genes. Instead of relying solely on educational programs to improve reading skills, we could intervene to enhance abilities at the genetic level. In one or two generations, we might be able to end the cycle of failure. The children, their families, and society would benefit. It is true that social and economic deprivation often plays a major role in fostering social failure.[15] But social conditions are notoriously hard to change. There may be room here for well-designed genetic interventions as well.

Similar enhancements could help internationally. In terms of health, we have spoken about improved resistance to infectious diseases. Here, gene vaccines could assist some poor nations afflicted with HIV/AIDS or malaria to escape a cycle of illness-induced economic backwardness and step into the modern world. Cognitive enhancements might also play a role. Reading skills and mathematical or computational abilities could speed development. By direct enhancement of a population's cognitive skills it might be possible to accelerate economic advances.[16] None of this should replace effective educational or social programs, but gene enhancements may be a useful supplement to whatever else is done, nationally or globally, to reduce inequalities and promote development.

Writing about the impact of science on our thinking about social justice, the philosopher Allen Buchanan observes that matters that one generation regarded as beyond their control often come under their descendants' direction, raising new questions of moral agency and responsibility. "The boundary between the natural and the social, and between the realm of fortune and that of justice is not static. What we have taken to be moral progress has often consisted in pushing back the frontiers of the natural, in bringing within the sphere of social control, and thereby within the domain of justice, what was previously regarded as the natural, and as merely a matter of good or ill fortune."[17] As Rawls's work indicates, what we regarded as unchangeably natural just a few decades ago is now becoming a matter of choice. In the name of social justice, we already intervene with enhanced schooling or other social programs aimed at improving the prospects of the least well-off members of society. Is it time to start thinking about gene enhancements as within the realm of our moral responsibility? Far from undermining our commitment to social justice, gene enhancements may be a tool for implementing it.

The core insight informing the Rawlsian perspective on this issue is that no one "deserves" their winnings in nature's genetic lottery, so we must be prepared to think about providing everyone access to an improved genetic endowment. But how likely is this to occur? In general, people are reluctant to give up their advantages or share with others in any way, or even to think from the perspective of the original position. This is a problem that plagues all theory in the field of political philosophy. In addition, there is a special problem connected to genetics. It is the problem identified by Sandel: that parental gene interventions may work to reduce our sense of a common fate and may feed our pride, making us unwilling to share anything with the less fortunate. Certainly, Roger Camden and his Yagaiist philosophy reflect this view. People like Camden use their head start in life to improve their children's lot, and they justify everything that follows in terms of a selfish philosophy of entitlement.

Is this an accurate depiction of people's likely thinking in an era of enhanced genetic control? I'm not sure. Recall that Sandel believes that shared vulnerability prompts us to feel solidarity with others. The example he offers is health insurance. Knowing that I could fall ill at any moment gives me an incentive to join with others in insurance schemes that protect us all. Sandel fears that improved genetic knowledge could erode such communal thinking. If genetic diagnosis gives me knowledge of my relative invulnerability to cancer or Parkinson's disease, I may choose to exempt myself from an insurance scheme, save the money, and manage on my own. Similarly, Sandel argues, the knowledge that I have "better" genes, or that I have passed "better" genes on to my children, could lead me to think that I am exempt from the shared human condition. I could come to regard myself as better than others, and abandon any solidarity I feel with their sufferings or misfortune.

But a moment's reflection shows us that Sandel's analogy is misleading. Health care is not like life. However healthy and fit I may now be, I always know that I remain highly susceptible to injury and illness. This prompts me to identify with others and to seek protection against reverses by pooling resources in insurance. But in life, just the opposite occurs. Fortunate and prosperous people usually ignore the contingency of their achievements. They regard themselves as self-made, and as long as they can get away with it, they resist efforts to make common cause with others. Rather than seeing themselves as merely lucky, they take credit for everything they have accomplished. They usually chalk up their achievements to hard work rather than to good fortune. And they take credit for their genetic endowment—in, for example, their lineage's good health or stamina. In some cases, they also attribute their advantages to divine favor, adding a powerful form of religious entitlement to their self-justification.[18] In other words, the state of mind that Sandel so fears *already prevails* in connection with genetics. People typically use their genetic good fortune to harden their hearts to others.

But if this is so, won't gene enhancements only worsen the

problem? Isn't Roger Camden the unavoidable result of this way of thinking? Perhaps. The problem is real. But my excursion into Rawlsian philosophy suggests another possibility. The availability of gene enhancements might undermine the sense that people somehow merit or have earned their start in life. We might all see much more clearly that many of our gifts are just that, not achievements or entitlements but the result of good fortune and biochemical luck. Furthermore, once people see that they can improve their own or their children's trajectory in life by genetic interventions, they may come to regard these as a universal entitlement. Just as we now regard access to prenatal care as something that all women should receive regardless of their economic circumstances, so we might come to think of a solid genetic endowment as a basic human right. The consequence of this way of thinking is to erode people's claim to ownership of good genes, to erase the sense that one's genetic endowment is beyond anyone's control, and to make us all regard sound, genetically influenced abilities as something to which we all have a right. The science journalist Nicholas Wade puts this matter succinctly when he observes, "The sequencing of the human genome makes it possible to envisage for the first time the creation of a genetically more just society, one in which the most fundamental kind of wealth—the genes that confer health and fitness—would for the first time be accessible to all."[19]

In the first chapter, I discussed this possibility in connection with sports. We saw that people tend to label gene doping "unfair" and, at the same time, entirely ignore the role of a good genetic endowment in athletic accomplishments. Julian Savulescu and his colleagues, basing their argument on the sheer arbitrariness of athletic endowment, make a case for permitting gene doping. They maintain that gene enhancements could allow many more people access to high-level performance and make athletics more egalitarian. The power of this argument lies in the way it reveals and combats the inherent genetic unfairness that already underlies our sports status quo. Not only is sports an arena of hard work and achievement, but it is skewed by people's genetic advantages. In this sense, it mimics life as a whole and opens up the prospect that gene enhancements might not worsen differences but help reduce them. Nevertheless, sports doping is hard to defend, because the aim usually is not to improve the game for all but to help one person win. Everyone else in a sport is made worse off when players pursue this kind of positional advantage. In contrast, gene enhancements in society do not have to aim at making one person better than others. By improving everyone's abilities in the performance of socially useful skills, they can represent a real benefit for society. Here, an equality of greater talent is not a liability but a social asset.

Which dynamics and which psychology will prevail in a world of gene enhancements? Will it be Roger Camden's Ayn Rand–like philosophy that says "I'm entitled to everything I achieve, including the genes I've received, and I don't have to share anything with anyone"? Or will it be a more Rawlsian view that says "My talents and my sound genetic endowment are not my doing but gifts that I haven't earned. I should regard them as social assets that should be used for the common advantage"? Will it be Roger Camden or Leisha Camden—entitlement or solidarity? The jury is out on the answer to these questions. But the answer is by no means as foreclosed as doomsayers like Fukuyama and Sandel believe.

People do take pride in their genes, which raises a further question of social justice: Will gene enhancements lead to the belief that some genotypes are "better" than others and that "inferior" genotypes should be eliminated or weeded out of the gene pool? Will it lead to a resurgence of eugenics? Could it promote coercive interventions by governments intending to "upgrade" their nation's genetic stock? All this happened in the United States, Great Britain, and Germany during the 1920s, 1930s, and 1940s. In the first three decades of the twentieth century, well before the rise of Nazism, the eugenics movement flourished in the United States and Britain and was responsible for legislation in twenty-eight American states requiring the sterilization of the genetically "unfit."[20] Under these laws, thousands of people too poor to defend themselves were institutionalized and surgically sterilized.[21] During the 1930s and 1940s in Nazi Germany, a philosophy of Aryan genetic superiority led to the social exclusion and sterilization of thousands of mentally handicapped people and ultimately to the murder of millions of Jews, Roma (often called Gypsies), and others branded as "subhuman" by Nazi racist ideology. Given this terrible history, the fear is that a culture or society committed to genetic "betterment" could once again unleash eugenic madness and provoke massive cruelties and injustices.

We should never underestimate the power of genetics to provoke injustice. Genetics seems to stir primal emotions in human beings. Since inheritance plays such an important role in shaping observable human traits, genes easily come to be regarded as the most basic component of a person's identity. In some ways, genes play the same role as the more traditional idea of the soul. Like the soul in religious belief, they predate our physical embodiment, and they persist through all the superficial changes of our life. Genes powerfully influence our observable bodily form.[22] Above all, genes relate to human bloodlines and families, determining with whom we can or cannot enter into intimate relations, who is related and who is not. As such, genes can easily be used as a basis for group inclusion—or exclusion. It is not surprising that people have been obsessed with genetic "purity" and "impurity" and that ideas like "genetic disease," "genetic fitness," and "gene superiority" have frequently become tools for oppression.[23]

Before we conclude that a commitment to gene enhancement must lead to disaster, however, it is important to recognize that the worst eugenics abuses of the past stemmed from at least three intersecting factors. One was the use of state power and coercion to enforce eugenic ideals. First in this country and then to an unprecedented degree in Nazi Germany, eugenics went from being an ideology to being an instrument of oppression when governments took over the job of enforcing genetic norms. In Germany, public health ideas that had been developed in the struggles against infectious disease were used to support coercive state interventions into familial decision making, including decisions about reproduction and about whom one could marry. Just as quarantine or compulsory vaccinations could be used to control contagion, so radical restrictions on reproductive liberty could be imposed for the sake of a nation's "genetic health."

A second factor contributing to the eugenic slide into tyranny was the bad, often harebrained science that underlay most eugenics. Eugenicists traced just about any undesirable social phenomenon or feature of behavior, character, or temperament to single inherited factors in the "germ plasm." In doing so, they ignored the fact that many traits are the result of environment, gene-environment interactions, or multiple gene interactions. Any trait that ran in families was assumed to be inherited.[24] Some examples of this radical simplification and confusion are amusing, like a study published in 1919 by Charles Davenport, head of the Eugenics Record Office and one of the leaders of the American eugenics movement, that claimed to have identified an inherited trait predisposing men to become naval officers. Davenport coined the term "thalassophilia" to describe this inherited "love of the sea." He believed that it was a "sex limited" (X-linked) trait found only in men. Of course, he failed to take into account two circumstances: that sons of naval officers often grew up in a seafaring environment and that women were traditionally barred from the navy.[25] If thalassophilia is merely funny, the eugenicists' tendency to trace complex behavioral traits and socially undesirable and antisocial ones to genes was catastrophic. Feeblemindedness, drunkenness, vagrancy, illegitimacy, prostitution, and poverty were all regarded as inherited conditions that could be stamped out by sterilizing those who exhibited them.

The third distinctive feature that made eugenics so dangerous was the elitist and often racist assumptions that went along with it. Both in this country and in Germany, eugenics became a tool for limiting the reproduction of lower-class citizens, "outsider" immigrant groups, and racial minorities, all of whose flaws were traced to defective genes. In my own state of Vermont, for example, small-town and urban middle- and upper-class groups took up the eugenics cause as a way of expressing their dislike and fear of Franco-American immigrants from Quebec and a threatening "hill country" population composed of poor people and people with some Indian blood.[26]

Once we see the special factors that made eugenics so dangerous, we have to ask whether a commitment to gene enhancement is likely to resurrect this very special combination of government oppression, perverted science, and social snobbery. There are reasons to believe a recurrence is possible. As I have said, the association of genes with volatile issues like family, blood, and purity means that there is always the risk that even good genetic science will be misused for bad social purposes. In addition, as the historian Diane B. Paul reminds us, the eugenics movement comprised many diverse and often contrary directions, from racially motivated state oppression to efforts to empower families. It even included high-minded and idealistic impulses to social reform. The attempt to avoid one kind of danger, Paul observes, may only invite another.[27]

Nevertheless, eugenics was a product of its time. In changed circumstances, gene enhancement may or may not lead to repressive eugenics.[28] For one thing, since the eugenic era, a strong commitment to individual reproductive liberty has become a major feature of human rights thinking across the globe. In the United States a series of judicial rulings and the rise of the feminist movement have made the right to individual reproductive decision making almost sacrosanct.[29] For another, in the post–Human Genome Project era, genetic science has entered a new phase. The tools are at hand to test claims of genetic causation, and a large and growing community of molecular biologists and geneticists stands ready to poke holes in careless or sloppy eugenics arguments. These counterweights, and the vivid memory of eugenic abuses, are real protections against the resurgence of the bad eugenics of the early twentieth century.

Some people believe that serious eugenics abuses can occur even without government involvement. The sociologist Troy Duster has introduced the phrase "backdoor to eugenics" to describe the growing use of state-mandated genetic screening programs for conditions like sickle cell disease. Although Duster feels that these programs can help reduce the incidence of some serious genetic disorders, he also believes that their introduction with little debate or public awareness is an invitation to more coercive state interventions in people's reproductive lives. [30]

Others worry that a host of nongovernmental actors could pressure parents to intervene to avoid genetic disease or to conform to some social standard of perfect health. In *Better than Well,* Carl Elliott notes that even those Americans who deplore the commercial or cultural forces that stimulate the desire for enhancements like straighter teeth and greater height still feel pressured not to deny these advantages to their children.[31] Writing in the same vein, Diane Paul argues that physicians seeking to avoid lawsuits, insurers trying to reduce their health care costs, and biotechnology firms interested in introducing new genetics products or services could work together to pressure parents into genetic interventions.[32] To Robert Wright the real threat is not a government program to breed better babies. "The more likely danger," he concludes, "is roughly the opposite; it isn't that the government will get involved in reproductive choice, but that it won't. It is when left to the free market that the fruits of genome research are most assuredly rotten."[33]

Some people worry about the impact on the future quality of life. Will parents, in the race to give their offspring competitive genetic advantages, ultimately produce a generation of Stepford children, millions of carbon copies made in our era's narrow conception of perfection? These children may fulfill today's popular ideal of the perfect child, but with their high IQs and Britney Spears / Brad Pitt good looks, won't their proliferation reduce the richness and diversity of human types? If so, future generations could be forever impoverished. Ultimately, this becomes a justice question of a different sort. Does one era have the right to impose its standards of human quality on every generation that follows? More than half a century ago in his book *The Abolition of Man*, C. S. Lewis put the problem this way:

> Each generation exercises power over its successors: and each, in so far as it modifies the environment bequeathed to it and rebels against tradition, resists and limits the power of its predecessors. This modifies the picture which is sometimes painted of a progressive emancipation from tradition and a progressive control of natural processes resulting in a continual increase of human power. In reality, of course, if any one age really attains, by eugenics and scientific education, the power to make its descendants what it pleases, all men who live after it are the patients of that power. They are weaker, not stronger: for though we may have put wonderful machines in their hands we have pre-ordained how they are to use them. . . . The real picture is that of one dominant age—let us suppose the hundredth century A.D.—which resists all previous ages most successfully and dominates all subsequent ages most irresistibly, and thus is the real master of the human species.[34]

Although I do not believe that eugenics in its bad twentieth-century form using state-sponsored coercion is likely to recur, the more subtle types of pressure that the critics fear—coercion from nongovernmental actors and self-coercion as parents race to keep up with the Joneses—could happen. But critics once again tend to ignore the presence in current practices of the things they worry about happening in the future, and they overestimate the likelihood of adverse future consequences. For example, to a small extent every human generation imposes its visions of biological perfection on the future. It does so through mate selection and social rewards that encourage some favored people to reproduce at higher rates than others. Evolutionary biologists have noted the role of sexual selection in molding the human body. A widespread male preference for females with large breasts and narrow waists, taken as signs of fecundity, seems to have influenced the shape of modern women.[35] In turn, the presence of substantial body hair on males, broad chests, and height may have occurred in response to females' sexual preferences.[36]

During the past half-century, by gathering together many of our brightest young men and women in elite colleges and universities, we have unintentionally embarked on what geneticists call a positive "assortative mating" experiment, in which breeding takes place among individuals with similar characteristics. Will the graduates of these schools, as they marry and have children, produce their own self-perpetuating intellectual elite?[37] Whatever the answer, the point is that in choosing our mates, we have always, consciously or unconsciously, somewhat shaped our children's inherited characteristics, and the choice of mate is not made free of peer pressure and the influence of prevailing norms.

The fear that future parents in pursuit of the perfect child will inflict millions of cookie-cutter children on the world is also overstated. Although people who must use donor sperm or eggs to have a child tend to choose donors with traits they believe will improve the child's chances of being attractive and intelligent, the governing reality is that most people want a child much like themselves (or maybe just a little bit better). In 1980 the inventor-eugenicist Robert Graham opened a "Nobel Prize winner sperm bank" that offered to couples where the male was infertile the option of artificial insemination with donor sperm from men who had demonstrated their high IQ.[38] Over the twenty years that it operated, Graham's "Repository for Germinal Choice" produced only 229 offspring in the United States and half a dozen other countries.[39] Not only did Graham overestimate infertile men's interest in having other men's genius children, he appears to have overestimated the clamor for perfection. As Teri Royal, director of one of the nation's largest surrogacy and gamete donor programs, observes, couples typically want a child whose intelligence is comparable to theirs and to whom they will be able to relate. "They're not looking for any Mensa applicant."[40] Here again, in other words, the natural dynamics of parenting offer protection against the critics' concerns. Parents may accept modest enhancements for their offspring, and they may try to avoid transmitting their own troubling blemishes, but the fear that they will be obsessed with the desire to have perfect children is probably groundless.

In sum, there are several reasons for thinking that a technology of gene enhancement will not lead to a massive and growing rift in society between the genetic haves and the genetic have-nots or to a frightening new eugenics movement. An awareness of the relative malleability of our genetic endowment may actually stir people to demand a better genetic start in life and reduce the differences in abilities that now sometimes perpetuate social failure. Improved science and a tradition of procreative liberty suggest that democratic societies are unlikely to drift back to the kind of statist control of reproduction that marked the eugenics movement. The tendencies of parental decision making provide us some confidence that the fears of genetic uniformity will not materialize. In my final chapter, I shall look at some of the social steps we can begin to take now to ensure that the good outcomes, not the dangerous ones, prevail.

One remaining justice issue is often raised: the concern that research on germline genetic research, even that directed at eradicating serious genetic diseases, could divert scarce resources away from more basic and necessary forms of medical research and care. Just a few years ago, a working group of the American Academy for the Advancement of Science (AAAS) spent a good deal of time studying the issue of inheritable gene modifications. Some of its members were particularly worried about this diversion of resources.[41] They pointed out that if the research is conducted with public funds to ensure proper regulatory oversight, the resources used for it will most likely come at the expense of other social investments, including research on disease conditions like malaria that affect far more people worldwide than do inherited genetic diseases. If the private sector takes over the research, not only could this lead to reduced oversight, but the resulting therapies are likely to be very expensive and available only to the well-to-do. Either way, the research will not benefit the economically less fortunate.

Given the present state of genetic science, it would be hard to advocate government funding for germline genetic research, especially for enhancement purposes. But it is not clear that anyone is either doing this or has to. Germline and enhancement work are both offshoots of genetic research in general. Research on muscular dystrophy and the muscle wasting associated with aging, for example, has had the unintended effect of opening up the possibility of gene doping in sports. Similarly, the Doogie Howser mice that I mentioned in chapter 3 illustrate the connections between basic genetic research aimed at understanding disease, on the one hand, and germline therapies and gene enhancements, on the other. It is not necessary to involve the government in germline or enhancement research. The presence of active research programs on the genetic underpinnings of serious diseases guarantees that discoveries will be made that open the door to germline gene enhancements.

The first applications of this research to human beings will probably take place in the private sector. At least initially, some private-sector resources will go to germline modifications. Apart from the larger question of whether this will eventually create a growing genetic-economic gap between the rich and the poor, these expenditures do not raise justice questions. We permit people to spend their money as they wish. If the wealthy can buy mountaintop mansions, luxury cars, and cosmetic surgery, why should they be prevented from spending their money on gene enhancements?

Eventually, society may have to step in to support research on germline interventions. This could come about, as the AAAS working group observes, to provide regulatory oversight or to break affluent people's monopoly of the technology and make it more widely available to all social groups.

Won't that investment represent a diversion of resources from more basic medical research or health and educational programs that benefit the poor? This is a complicated priority question for our descendants. But one thought is helpful: in some cases, gene enhancements may be the most efficient and *least costly* way of addressing urgent social problems.

Some years ago, the physician–health-care theorist Lewis Thomas introduced a distinction among three levels of medical technologies. The first was "non-technology." This took the form of "caring" or "standing by" and was usually applied to untreatable disease conditions like terminal cancer, severe rheumatoid arthritis, and advanced cirrhosis. The second was what Thomas called "halfway" technology. It includes advances that many regard as medicine's greatest achievements: organ transplantation, renal dialysis, respiratory and coronary intensive care units, coronary artery bypass or stent surgery, and cancer treatment involving surgery, radiation therapy, and chemotherapy. This kind of technology is enormously expensive. Thomas described it as at once "highly sophisticated and profoundly primitive. It is the kind of thing that one must continue to do until there is a genuine understanding of the mechanisms involved in disease."[42]

Finally, there is high technology that is powerful in terms of curing or preventing disease. Vaccination programs and the use of antibiotics to eradicate infectious diseases are illustrations. In Thomas's words,

> The real high technology of medicine . . . comes as a result of a genuine understanding of disease mechanisms, and when it becomes available it is relatively inexpensive, relatively simple, and relatively easy to deliver. . . . I cannot think of any important human disease for which medicine possesses the outright capacity to prevent or cure where the price of the technology is itself a major problem. The price is never as high as the cost of managing the same diseases during the earlier stages of non-technology or halfway technology. . . . It is when physicians are bogged down by incomplete technologies, by the innumerable things that they are obliged to do in medicine when they lack a clear understanding of disease mechanisms, that the deficiencies of the health-care system are most conspicuous. If I were a policymaker, interested in saving money for health care over the long haul, I would regard it as an act of high prudence to give priority to a lot more basic research in biologic science.[43]

Thomas's focus is on disease. This has direct application to germline genetic therapies that correct or prevent costly disorders. But his larger point about technology also has application to social and economic problems. Persistent problems in reading or mathematics and behavioral conditions that lead to social failure are currently either not addressed (non-technology) or are addressed by what amount to halfway social interventions (intensive remedial educational programs). Unaddressed, these problems incur great social costs. Addressed at the halfway level, the costs are still enormous (and are often given as a reason for not investing in them).

An illustration is the problem of iodine deficiency. About two billion people—a third of the world's population—get too little iodine, including hundreds of millions of people in India and China. Studies show that iodine deficiency is the leading preventable cause of mental retardation. Even a moderate deficiency in pregnant women and infants causes brain damage and lowers a child's intelligence by 10 to 15 IQ points, shaving incalculable potential off a nation's development. Trying to rectify this physiological damage after it occurs is nearly impossible, and even reducing its social impact through educational interventions is prohibitively expensive. In contrast, recent global efforts to address the problem by the simple measure of iodizing salt make a significant contribution to helping nations escape the cycle of poverty.[44]

If efficient gene enhancements could also help reduce some health and medical problems, they might prove to be very cost effective. In addition, as Allen Buchanan has pointed out, gene interventions have some ethical advantages over halfway social interventions. They "might in fact be less threatening to individual autonomy and privacy and more likely to succeed at removing barriers to opportunity than the kinds of large-scale changes in family life needed to counteract the opportunity-limiting effects of disadvantages in family culture and early childhood experiences."[45] So the choice between expenditures on social needs and expenditures on gene research may be a false one. Expenditures in one direction need not detract from expenditures in the other. The quest for genetic advances should never replace attention to the serious health care needs of so many poor people and poor nations. But genetic technologies could also offer new, imaginative ways of addressing not only medical problems but social problems as well.

The mention of addressing social problems through genetics unavoidably brings up eugenics. The question remains, therefore, whether gene modification will ease social conflict or whether it will lead to horrific new abuses. Will it produce the world of Eloi and Morlocks that H. G. Wells depicted? Will it cause the near catastrophes of Nancy Kress's *Beggars* trilogy? Or will it help the human species attain new levels of equality and social justice? Some of the answers to these questions lie in the details of implementation. *How* we move into an era of gene modification will profoundly shape the direction that era takes. In the final chapter, I shall look at some ways we can monitor and modulate our movement forward and increase our chances of success. For the time being, my aim is to disturb our certainties. People have seen gene enhancement technology as only a threat to society and social justice. The threat may be real. But I have tried to start us thinking about the possibility that genetics may have a positive role to play in our dreams of a more just world.

CHAPTER *7*

Playing God

From a seat on a cloud, the gray-bearded deity hurls lightning bolts down on the earth below, but each bolt is shaped like a double helix. The message of this cartoon, which appeared in one of the earliest federal reports on genetic engineering, is clear: if human beings try to usurp God's powers, they will feel his wrath.[1]

We may smile, but for many people the risk of provoking God's anger by genetic engineering is anything but funny. They believe that gene manipulations represent human pride and arrogance at its worst, threatening to put human beings in direct opposition to God's will and risking the gravest consequences for individuals and society. Thirty years ago, Paul Ramsey, a conservative Christian bioethicist and one of the first scholars to address this issue, sounded a warning in his book *Fabricated Man:* "Men ought not to play God before they learn to be men, and after they have learned to be men they will not play God."[2]

How can we understand this worry about playing God? At first sight, it seems to have considerable force. The God of the Bible is uniquely associated with the design and creation of living things. In the first two chapters of the book of Genesis, God brings plants and animals into being from nothingness. The first human being, Adam, is created from the dust of the ground when God breathes life into his nostrils. According to a long tradition in the Jewish, Christian, and Muslim faiths, this original divine creative act repeats itself every time a new human being comes into existence. Parents are agents of God, but it is God who starts each human life by breathing a soul into the developing human being at some point between conception and birth. For those influenced by these traditions, human control of genetic makeup represents a Promethean seizure of God's power. Some, like Leon Kass, the former head of President Bush's bioethics council, regard genetic interventions as humankind's contemporary replay of the Tower of Babel episode.[3] They see it as the effort by limited and puny creatures to seize control of the one domain that is God's alone. Many people agree. Public opinion polls conducted since the beginning of the era of genetic engineering in the early 1980s reveal that nearly two-thirds of Americans believe that altering human genes is against God's will.[4] In a nationwide telephone survey commissioned by *Harper's Magazine* in 1997, people were asked, "If you had to choose one of the following, who should have the power to control the genetically linked characteristics of a child before birth?" Fewer than 1 percent of the respondents chose "the doctor"; only 11 percent chose "the parents"; 16 percent chose "no one"; and fully 70 percent chose "God."[5] This strong current of religious thinking continues to shape Americans' response to genetic interventions. Even those who don't express their views in religious terms tend to share these concerns.

Nevertheless, when we place gene technology in the context of other human scientific and technical achievements, the force of these objections becomes less compelling. It is hard to see just what the prohibition on "playing God" means and whether it makes any sense. For example, the same Genesis narratives that many read as a source of the prohibition confer on human beings the task of governing and tending to nature. Throughout the Bible, agriculture, animal husbandry, metalworking, and many other technological interventions in nature are permitted and even approved. Medical interventions in our own bodies are also permitted. Although the Bible sees God as the ultimate source of healing, it takes for granted that human beings can intervene to cure diseases and injuries. Judaism, Christianity, and Islam have always approved of medical care, praising medical professionals as God's agents on earth.

Is it genes, then, that are off limits? Is the problem that while ordinary, medical interventions, including surgery, are permitted, genes are too close to God's unique area of control? In fact, almost no religious thinker who has offered an opinion about human genetic modification draws the line here. A widely shared assumption among religious commentators is that it is morally permissible to use genetic interventions for the purpose of curing disease. Pope John Paul II, who took a very conservative position on many new bioethical questions, repeatedly approved of gene therapies so long as they respected other Catholic reproductive norms (such as the bans on contraception and in vitro fertilization) and were aimed at curing or preventing disease. Speaking in 1982 at the Pontifical Academy of Sciences, the pope observed that modern biological research "can ameliorate the condition of those who are affected by chromosomic diseases," and he lauded this as helping to cure "the smallest and weakest of human beings . . . during their intrauterine life or in the period immediately after birth."[6] A year later, addressing a meeting of the World Medical Association, the pope repeated this view. A strictly therapeutic genetic intervention, he said, "will be considered in principle as desirable, provided that it tends to real promotion of the personal well-being of man, without harming his integrity or worsening his life conditions."[7] In these and other remarks, the pope did not distinguish between somatic and germline interventions, but in 2003, a theological commission headed by Cardinal Joseph Ratzinger (now Pope Benedict XVI) explicitly approved germline genetic engineering involving genetic alterations in a man's sperm-producing cells.[8] Although other Christian commentators have sometimes urged special caution where germline interventions are involved, the consensus of opinion is that interventions aimed at preventing or curing disease, including those that safely eliminate a genetic disorder from a family line, are consistent with the traditional permission for human medical care. In the words of John Paul II, a curative intervention, whether genetic or otherwise, generally "falls within the logic of the Christian moral tradition."[9]

The concern about playing God, then, seems to narrow down to one thing: gene enhancement. What most bothers religiously inclined people who object to playing God are interventions that aim at changing or improving human nature and that seek to provide individuals with enhanced abilities. Here the concern about playing God attains its greatest force. John Paul II expressed this view when, in the same discourse in which he approved of medical genetic interventions, he urged avoidance of "manipulations tending to modify the [human] genetic store."[10]

But why are genetic cures acceptable, whereas gene enhancements are not? In *What Sort of People Should There Be?* the British philosopher Jonathan Glover asks whether the concept of playing God makes sense and whether it helps explain the distinction between permissible and impermissible gene interventions. Glover notes that religious communities already allow many types of genetically related interventions. He adds that it is odd to permit gene therapies aimed at curing disease (what he calls negative genetic engineering) while banning enhancements (positive engineering): "The prohibition on playing God is obscure. If it tells us not to interfere with natural selection at all, this rules out medicine, and most other environmental and social changes. If it only forbids interference with natural selection by the direct alteration of genes, this rules out negative as well as positive genetic engineering. If these interpretations are too restrictive, the ban on positive engineering seems to need some explanation. If we can make positive changes at the environmental level, and negative changes at the genetic level, why should we not make positive changes at the genetic level? What makes this policy, but not the others, objectionably God-like?"[11]

There is a reply to Glover's questions, but one that raises even more questions. The reply takes us back to the opening chapters of Genesis, where we read that God repeatedly declares the creation, including the creation of human beings, to be "good" or "very good" (Genesis 1: 4, 10, 12, 18, 25, 31). Later in Genesis we learn that Adam and Eve, in an effort to achieve divine omniscience, disobey God's command not to eat of the tree of the knowledge of good and evil, and they suffer a host of punishments. These include eviction from the Garden of Eden, difficulty in cultivating the earth for food, the pain associated with childbirth, and the inevitability of aging and death. In later Jewish, Christian, and Muslim thought, all these departures from the blessedness that Adam, Eve, and their offspring would have experienced in Eden are the result of sin. These traditions also proffer the hope that a return to God and the attainment of salvation could restore creation to its primal integrity and goodness.

These ideas help make sense of many religious people's comfort with curative medical interventions, even genetic ones. They see them as aimed at reversing the ravages of sin. In our roles as co-creators and obedient creatures on the path to redemption, we have a limited authority to reverse the corruption of the created order and our physical bodies that Adam and Eve's transgression began. But in a world created by God as "good," this authority does not extend to making fundamental changes or "improvements." The world came into being with divine blessing, and efforts to change it—other than to repair the damage we ourselves wrought—are seen as a further rejection of God's sovereignty and an impermissible playing God. This view applies as well to interventions designed to genetically enhance a child's capabilities. The child's genome, like the human genome generally, comes into being at God's behest. Although the correction of disorders can fit within our general permission to reduce the ravages of sin, gene enhancement amounts to second-guessing God's will for that child and inserting ourselves in God's place.

I have spelled out this complex, Genesis-derived view because I think it underlies and explains a great deal of the religious opposition to nondisease-related genetic interventions. To this way of thinking, gene enhancements amount to defying God's will for the created order. I have repeatedly observed that human beings are prone to status quo bias, overestimating the value of existing conditions and conjuring up reasons why changing them could be catastrophic. What we are seeing here is that our religious traditions give this whole way of thinking an aura of sanctity. The human genome, in its present form, is not just good: it is divinely created. Tampering with it is more than unwise: it is sinful.

Once we identify this powerful idea, it becomes reasonable to ask whether it is right. Does the genome have the sanctity that this reading of biblical faith suggests? I ask this question in a friendly way. Many people do not accept biblical teachings and do not care what the Bible says about these issues. But we can try to take biblical faith seriously and ask whether, within its confines, it really is the case that Creation must be regarded as perfect at its origin and that the uncorrupted human body was never meant to be subject to further refinement. In fact, the answer to this question is not clear, but that is noteworthy. What seemed obvious before we asked the question becomes far less so once we examine it with critical eyes.

The Protestant ethicist Ted Peters believes that the Bible offers no warrant for believing the Creation was meant to be fixed and unchangeable. Speaking from within the monotheistic Christian tradition, Peters asks, "What is sacred: God or the Creation?" If the answer is God, then the natural world cannot be made into an untouchable holy reality. Nature is not its own author and is not an absolute. It depends on and is subordinate to a divine creator who transcends it. Quoting the conservative Christian biomedical ethicist Hessel Bouma III and his colleagues at the Calvin (College) Center for Christian Scholarship, Peters adds, "To presume that human technological intervention violates God's rule is to worship Mother Nature, not the creator." For Peters, Bouma, and others, human beings' role as God's co-creators gives us the power and authority to intervene in nature and improve it. This capacity extends to our genes. "To nominate DNA for election into the halls of functional sacredness," says Peters, "is arbitrary."[12]

Some Jewish thinkers point to explicit teachings in their tradition that authorize humans to attempt post-Creation bodily improvements. The Talmud records a debate among several great rabbis about the limits of God's creative activity. The specific topic under discussion is circumcision. One of the rabbis points out that God created wheat but not bread; milk but not cheese. The same is true for the human body. Human males are not born pre-circumcised. Although God created the human body, he enjoined some post-Creation "improvements," of which circumcision is one. Drawing on these rabbinic debates, the conservative Jewish bioethicist Rabbi David Bleich states that there is "no evidence either from Scripture or from the rabbinic writings that forms of intervention or manipulation not expressly banned are contrary to the spirit of [Jewish] law." The opposite is true. "Jewish tradition, though it certainly recognizes divine proprietorship of the universe, nevertheless gratefully acknowledges that while 'the heavens are the heavens of God' yet 'the earth has He given to the sons of man' (Psalms 115: 16)."[13]

These Jewish teachings have particular resonance today. New evidence suggests that circumcision may be highly protective against sexually transmitted diseases. In late 2006 an NIH random-controlled clinical trial conducted in Kenya and Uganda was abruptly halted when it revealed that circumcision reduced men's risk of HIV infection by more than half. Those responsible for the study concluded that the benefits of circumcision could not ethically be withheld from the uncircumcised control group.[14] Here we have a case where medical knowledge and ancient religious practices both encourage a modification of the human body. But if we can change the body in this way, why should our interventions not extend to the genes? And if religion permits alterations to control disease, why not positive enhancements?

In short, some religious opposition to gene enhancement appears to rest on a view of the created "goodness" and finality of the (healthy) human genome, a view for which there is neither clear warrant nor a theological consensus. As is so often the case, status quo bias (in its religious and other forms) appeals to our emotions, but it does not always withstand close scrutiny.

Although a major source of opposition to gene enhancements is the religious argument based on an objection to playing God, this viewpoint is also frequently buttressed with arguments drawn from science. On many new issues of bioethics, like stem cells and cloning, conservative religious opinion and science-based views move in opposite directions. But when it comes to gene modification and enhancement, religion and science often join hands, reinforcing the intensity of the opposition.

Some science-based opposition to human gene enhancement draws on ideas developed by the environmental movement. A key belief is that natural ecosystems have a stability and durability that they acquire over thousands or millions of years of evolution. When human beings disturb these systems to satisfy shortsighted human needs, they can destabilize the systems and cause disaster. The experience with the use of the pesticide DDT is an example. Aimed at improving crop yields, it ended up creating DDT-resistant insects and devastating bird populations when it destroyed the female birds' ability to form sturdy eggs. For some scientists and environmentalists, the human genome is an ecosystem like this. It has evolved over millions of years, with natural selection continually weeding out dysfunctional genes and ensuring that individuals who survive are optimally suited to their environment. There is a "wisdom" to evolution that confers an integrity on the human genome that we alter at our peril.[15] Some advocate regarding the human gene pool as a "genetic patrimony," a treasured resource that we must preserve intact for future generations.[16] Others have gone so far as to propose an international "Convention of the Preservation of the Human Species" that would outlaw all efforts at inheritable human gene modifications.[17] In the late 1980s the German Bundestag's Commission of Inquiry into Genetic Engineering declared that any change of the human germ line constituted "an act against nature."[18]

People who view the human genome as a once-and-for-all perfect creation by God and those who view it as the result of millions of years of successful Darwinian evolution tend to have little in common. Evolutionists and creationists are usually locked in mortal combat. Yet on this issue, the two perspectives unite, and it is a union with powerful implications. It means that anyone advocating human gene modifications is criticized from the cultural right and the cultural left. With the vast number of people in the middle either uninformed or undecided, the union creates a hostile environment for defending the possibility of gene modification.

I am not just speculating about the union of opposing ideas. A few years ago, as the President's Council on Bioethics prepared to write its report on biomedical enhancements, it invited testimony from a host of researchers in the field. One of the speakers was H. Lee Sweeney, the University of Pennsylvania researcher whom I mentioned in chapter 1. Sweeney provided the council with an overview of his research on muscle-enhancing somatic cell gene therapy in mice. After Sweeney made his presentation, a council member, William Hurlbut, a physician with an interest in bioethics, questioned him. Like many of the other Bush appointees to the council, Hurlbut is uncomfortable with embryonic stem cell research and related innovations in reproductive medicine. He also brings a conservative, religiously influenced viewpoint to the enhancement issue. Here is how the transcript of that day's hearings records the exchange between Hurlbut and Sweeney (the italics are mine).

DR. HURLBUT: You haven't seen any insertional mutagenesis or anything like that? And, by the way, I want to ask you also: have you seen an increase in longevity of these mice?

DR. SWEENEY: We've seen a slight increase in longevity, and it's small enough that we haven't got enough animals yet to make it statistically significant, but it might be ten percent or so at the end of the day. . . .

DR. HURLBUT: So is basically your answer that, to sum it up, you see this as a therapeutic agent in the face of a difficulty, but it would otherwise be very reckless at least in the near term and maybe even in the long term to do this kind of thing?

DR. SWEENEY: Yeah. Well, how reckless it is is going to have to be determined by a lot of safety testing that has yet to be done, and so it's difficult for me to see how reckless it would be in the long term. I mean, my personal bias is that the general approach, if it can be technically made more—if it can be done more easily, I think would achieve our goal of in the aging population or in a dystrophic population of in one case improving quality of life and in the other case extending life. . . .

DR. HURLBUT: Well, that's a hard comment to follow because what I was going to say was not just with a feeling of life changed from within a person, but the feeling of the relationships toward that life would change.

I mean, I remember watching my own father go through significant muscle atrophy and taking on a whole new relationship with him as I helped him move, and that was a meaningful part of the end of my relationship with my father. And so I think in a way the larger question comes overarching here. *Is the world in some way good either by the benevolent purposes of a creator or by the harmonious balance of a subtle evolutionary force or both, or is it just that one function was preserved because it helped the organism leave its genes in the next generation? . . .*

DR. SWEENEY: Yeah, I guess it comes to a more philosophical issue. Was the aging process by design or simply neglect? . . . I mean, is it a designed process or is it just the lack of forethought at a time in life when you're no longer contributing to the propagation of life? . . .

DR. HURLBUT: Have you heard the saying Mother Nature always bats in the bottom of the ninth? . . .

DR. SWEENEY: Well, again, I think it's whether or not Mother Nature designed the aging process or really just it's a fallout of not caring about the process.[19]

Hurlbut's question "Is the world in some way good either by the benevolent purposes of a creator or by the harmonious balance of a subtle evolutionary force or both?" encapsulates the viewpoint I am signaling: both evolution *and* God have conspired to perfect the human genome. What is remarkable about this exchange is that Sweeney, the scientist, is reluctant to buy either proposition. When Hurlbut tries to defend the inherent goodness and naturalness of the human aging process, Sweeney pushes back. He points out that attempts to understand aging in terms of evolution may fail because natural selection doesn't operate during the postreproductive years. The diseases and morbidity of aging may be accidental results of other biological processes—what Sweeney calls nature's "neglect" or "lack of forethought" for a time of life no longer under selective pressure.[20]

A good deal of scientific evidence supports Sweeney's hesitancy. Besides aging, many features of the human genome may be accidents of biology that cannot be explained in terms of how they contributed to human evolutionary survival or functioning. In chapter 4, I mentioned the discovery of a gene that contributes to aging by rapidly destroying cells that show any signs of malfunctioning from oxidative stress. The normal hair-trigger setting of this gene works well in younger individuals who have plenty of spare cells, but it hastens aging in older individuals. In one study, an Italian team of researchers created genetically engineered mice in which this gene was disabled. The mice lived 30 percent longer than normal mice did, with no apparent harm.[21] Here we find a biological function designed to assist survival and reproduction in younger animals but one that continues to operate, with unfortunate results, long after it makes its contribution. This hair-trigger mechanism has not been eliminated from the genome by natural selection, because its harmful impact in old age had no negative effect on reproduction.

Imperfections in nature also sometimes result from the cumbersome way that evolution proceeds, building one small advance on another and incorporating rather than rejecting outmoded elements dating from an earlier stage of development. This process produces biological systems that are not always optimally designed. In the human eye, for example, the retina is inside out. The nerve fibers that carry the signals from the eye's color- and light-sensing cells lie on top of the cells; to get to the brain they must plunge through a large hole in the retina, creating a blind spot in our vision. As Daniel Dennett comments, "No intelligent designer would put such a clumsy arrangement in a camcorder, and this is just one of hundreds of accidents frozen in evolutionary history that confirm the mindlessness of the historical process."[22] Ever since the early nineteenth century, when Bishop William Paley first made the point, the amazingly complex structure of the eye has been a mainstay of arguments to prove the existence of an intelligent designer of the universe. A close examination of the eye's structures not only weakens the argument but shows that evolution produces far from perfect results.

Finally, even when evolution works to produce organisms ideally suited to survive and reproduce in their environment, the organism's capacities are limited to that environment. If the environment changes, the organism's traits may no longer prove advantageous. Human beings evolved as highly successful hunter-gatherers on the African savannah and quickly spread around the world. But the African savannah of a hundred thousand years ago was very different from today's teeming, energized, urbanized world. Aspects of human biology that were highly adaptive back then may be far less so today. This includes such traits as our propensity to overeat and store fat in almost all circumstances.[23] This tendency had survival value when famine was constantly threatening, but today, in an era of readily available food and reduced physical activity, it contributes to the epidemics of obesity and diabetes sweeping almost every country. The number of people around the world suffering from diabetes, in developing as well as developed countries, has skyrocketed in the past two decades from 30 million to 230 million.[24] Similarly, our intense tribalism, the powerful tendency to bond with relatives and friends and scorn or reject outsiders, had obvious benefits when humans lived in small, competing bands. But the same behavioral tendency now provokes violence and genocide. In human beings, evolution has produced a remarkable species, but there is no scientific reason for believing that our biological nature is beyond improvement.

The belief that ours is a perfect genome endowed by God and nature makes it seem morally and spiritually dangerous to tamper with it. Once we realize that the genome is neither religiously sacrosanct, at least according to biblical faith, nor biologically perfect, we can ask whether genetic modifications might help us remedy not only physiological imperfections but also some of the serious moral and spiritual problems facing the world community. Is it possible that we might use genetic means to improve the moral performance of our species?

This question is asked by the philosopher Jonathan Glover, influenced by the biologist Richard Dawkins's idea of the "selfish gene." According to Dawkins, genes are essentially "selfish replicators" that persist in our genome because they were better able to survive than other DNA sequences. This means, says Glover, that "unless we have a very crude form of evolutionary ethic, there is no reason why the qualities we value should exactly coincide with those which have led to gene survival." It is possible, says Glover, that "the world would be a better place if people were more altruistic and generous than a perfectly calculated survival strategy for genes would make them." In Glover's view, the problems are not hard to identify: "We have the desire to identify with a group; willingness to hate other groups; obedience; conformity; aggression; the responses to being part of a crowd; the responses to stirring music, and to uniforms, flags and other symbols; the love of adventure and risk; the need for a simple framework of ideas to make sense of the world; and the desire to believe rather than doubt. These aspects of our nature are not all entirely bad, and they may not contribute equally to war. Some of them may be more easily modified or controlled by social changes than others. But, taken together, they have contributed to the catastrophes we all know. If either environmental or genetic methods are available, we may be wise to change ourselves if we can."[25]

Glover is not naive. Even if some of these human tendencies have a genetic contribution, they probably are not traceable to single genes, and they will probably not be susceptible to modification any time in the foreseeable future. Although there is growing evidence that many behavioral conditions, from alcoholism to addictive behaviors to aggression, have strong genetic underpinnings, behavioral genetics, the study of how genes contribute to complex human behaviors, temperament, and cognitive performance, is still one of the less advanced branches of genetic science.[26] It will be some years before we can confidently alter some of the tendencies that continue to mar human social life. Furthermore, even if we can identify genes underlying undesirable human behaviors, some of our bad propensities may be essential for human flourishing. If we reduce our tendency to be selfish on behalf of our group, will we also reduce the bonding and loyalty needed for important collective endeavors in science, education, business, and the arts? Finally, for spiritual or moral progress, it is hard to imagine a simple biochemical fix. As the biologist and Anglican priest Arthur Peacocke reminds us, "The biological evolution of man has now been superseded by [his] psycho-social development."[27] Human moral achievement involves enormous personal effort and cultural learning. It has taken us generations to see the foolishness of excessive nationalism. No modification of genes is likely to replace this growth in cultural wisdom and experience.

It is also true, however, that we seem to keep repeating our mistakes. Something keeps propelling us to disaster. The tribal instincts that once served us so well continue to aggravate modern problems. Years of research may be required to find ways of modifying our genes to produce a species even slightly more suited to the complexities of a global civilization than humans are now, but the possibility of doing so cannot be ruled out. Indeed, recent research on animal behavior points toward the existence of a suite of genes, shared across a variety of mammalian species, that determine whether an individual animal is skittish or friendly, tame or violently aggressive. Some believe that human beings have evolved a set of the milder genes in this suite as a consequence of our long history of social life, raising the possibility that we might someday be able to deliberately accelerate this evolutionary progress.[28] In short, gene modifications may have a place, alongside cultural efforts, in achieving the larger goal of human spiritual and moral fulfillment.

Philosophers like Jonathan Glover can identify problems and opportunities, but only artists can portray them. Although the overwhelming majority of fiction writers, from Aldous Huxley to Margaret Atwood and Ursula Le Guin, have depicted the disastrous biological, moral, and spiritual effects of gene modification, one writer offers a different vision. The African-American science fiction novelist (and MacArthur "genius" award winner) Octavia Butler provides us with an imaginative glimpse into a world where the existing genome has led to disaster and where only radical genetic change can save humankind. In this world, moreover, religious resistance to change is seen as contributing to the human dilemma.

Butler develops her vision in a series of three novels, *Dawn* (1987), *Adulthood Rites* (1988), and *Imago* (1989), known together as the *Xenogenesis* trilogy, or *Lilith's Brood*. Her feminist perspective and African-American background inspire the fictional future world created by Butler in the trilogy. Butler establishes her outlook in her first and most widely read novel, *Kindred* (1979). It tells the story of Dana, an African-American woman living in 1976 who, against her will, is repeatedly thrust back in time to the antebellum South. In some inexplicable way, she is summoned to ensure the survival of a young slave-owner's son, who turns out to be her ancestor. During her journeys back, Dana witnesses firsthand the violence and cruelty of slavery. She also repeatedly encounters the conservative, religiously based ideology that espouses white racial superiority and condemns any form of race mixing while turning a blind eye to the rape of black women by white slave masters. *Kindred* establishes the foundation of Butler's perspective in the *Xenogenesis* trilogy, one that sees appeals to genetic "purity" as a smokescreen for racial and sexual domination.

In the trilogy itself, Butler takes this theme into the future. The novels trace the career of Lilith Iyapo, a dark-skinned female survivor of a nuclear war on Earth that was followed by a nuclear winter wiping out most life on the planet. Lilith awakens, after several centuries of imposed suspended animation, on a giant spaceship of the Oankali people that is more like a living organism than a machine. The Oankali are an extraterrestrial trader race that travels the universe in search of species with whom they can exchange genetic material. As one of the Oankali tells Lilith, "We trade the essence of ourselves. Our genetic material for yours."[29]

Lilith learns that the Oankali require three sexual partners to reproduce: a male and female of the visited species (in this case, humans) and an ooloi, or skilled genetic engineer who is neither male nor female. The Oankali have suspended human beings' ability to reproduce on their own. This means that human continuance can be accomplished only through a blending of human and Oankali genes.

Although the Oankali have restored the surviving human beings to health, removing all hereditary diseases and cancer, they refuse to reactivate humans' ability to reproduce on their own. They are convinced that human beings possess a devastating inherited flaw that condemns them to destroy themselves and others. This flaw is a contradiction in their natures. They possess very high intelligence (they are among the most intelligent species the Oankali have met), a result of their recent evolution, but they also have an ancient tendency toward hierarchy and domination inherited from primate ancestors. Together, these traits lead human beings to ever higher levels of technological capability, but as the nuclear war shows, they ultimately use these traits to destroy.

The Oankali are natural genetic engineers. They intend to trade genes with human beings to produce new beings who incorporate humans' biological strengths, including their tendency to cancer, which, in ooloi hands (and in an anticipation of our current discussions about stem cell research), promises new approaches to cell growth and regeneration. But they also hope to leave behind humans' self-destructive weaknesses.

In *Dawn,* we meet Lilith when she has been selected as the trainer for a new generation of reawakened human beings who will be prepared in a shipboard simulator, the "training floor," for return to the terrestrial Amazonian environment to help repopulate the planet. Although the Oankali and ooloi in various forms initially seem physically repugnant (they have gray, highly manipulable sensory tentacles all over their bodies), Lilith reluctantly succumbs to their intellectual and sexual powers and agrees to play her assigned role. She enters into a sexual union with an ooloi named Nikanj and with Joseph Li-Chin Shing, a gentle, reawakened male. She is genetically enhanced to eliminate hereditary cancer and to improve her strength and longevity. However, she shares one feature with Lilith, the first wife of Adam, who, according to rabbinic lore, refused to obey her husband and was replaced by the more compliant Eve.[30] Like her namesake, Lilith Iyapo is reluctant to cooperate with her masters' creative plans.

Forty reawakened human beings are placed in Lilith's care, but they soon prove openly defiant. One of them, Curt, a tough former

New York City cop, resents the leadership of a woman. He and other men seek to enlist the human females in an effort to throw off Lilith's control. The males view the Oankali project as emasculating them—making them "female" to the technologically superior and sexually assertive Oankali—and they rebel. Two of them turn against Lilith's mate Joseph. One decides that he is something called a faggot; the other dislikes the shape of his eyes.

Violence flares, and Curt hacks Joseph to death. Because the Oankali abhor killing, including capital punishment, they return Curt to perpetual suspended animation and transport the remaining humans to Lo, the Earth settlement. Lilith learns that she is pregnant with a child resulting from the union with her Oankali tutors, the ooloi Nikanj, and Joseph, whose seed was saved by the Oankali for this purpose.

Lilith is never presented as an omniscient narrator but as one human being struggling with her circumstances. She remains deeply ambivalent about her mission. In a conversation with the ooloi Nikanj, Lilith questions the Oankali's plans for perpetuating human genes only in combination with their own. "Some will think the human species deserves at least a clean death," she says.[31]

Nikanj replies, "Our children will be better than either of us. We will moderate your hierarchical problems and you will lessen our physical limitations."[32]

"But they won't be human," Lilith says. "That's what matters. You can't understand, but that *is* what matters."[33]

To the very end of the novel, Lilith is unsure whether she is helping the human race survive or disappear. She wonders whether human beings could learn to have children on their own. "If she were lost, others did not have to be. Humanity did not have to be."[34]

In the second novel of the series, *Adulthood Rites,* the difficult tensions exposed on the spaceship continue on Earth. "Resister" communities arise among human survivors seeking to escape Oankali supervision. They regard Lilith as a traitor who is cooperating in the bastardization and erasure of their species. They preach a biblically based religious ideology that abhors the genetic "uncleanness" of human-Oankali merging. The men dominate the women and are quick to use primitive rifles to kill Oankali, human-Oankali "constructs" (genetically blended individuals), and one another.

The chief protagonist of this novel is Lilith's construct child, Akin, who matures into a young man able to appreciate both the Oankali and the human perspectives. As the novel ends, Akin persuades the Oankali to allow a small band of human beings to migrate to the cold and lifeless planet Mars, where their fertility will be restored and they can try to reconstruct their civilization. Nevertheless, although the Oankali yield to Akin's wishes, they resist this concession. It is not that the Oankali resent human continuance. They share Akin's wish for the well-being of the species, but they believe in the absolutely determinative power of humans' genes and are convinced that giving humans a false hope of redemption is cruel. The Oankali know that human beings will again destroy themselves. One of the Oankali tells Akin, "You and those who help you will give them the tools to create a civilization that will destroy itself as certainly as the pull of gravity will keep their new world in orbit around its sun."[35]

The final novel in the trilogy, *Imago,* follows the career of Jodahs Iyapo Leal Kaalnikanjlo, the construct child of two human parents (one of whom is Lilith) and the ooloi Nikanj. Oankali children normally take on the sex of the parent they are most drawn to. In Jodahs's case, this has turned out to be the ooloi Nikanj. Because of Jodahs's untested and possibly dangerous powers as a flawed natural genetic engineer, he is exiled to a remote region of the Andes, where he forges an alliance with a group of resisters willing to accept his guidance.

The novel—and the trilogy—ends on a note of hope and maternal creativity typical of the Oankali. Jodahs reaches into his yashi, a place beneath the sternum where ooloi keep a vast repository of genetic information, and finds a seed that can become a new city. This will grow on Earth and eventually become a spaceship that can journey to the stars. Jodahs plants the seed and says, "Seconds after I had expelled it, I felt it begin the tiny positioning movements of independent life."[36]

The perspective that Butler develops in these novels is dramatically opposed to that found in so many of the current ethical and religious discussions about human genetic engineering. The more common view is that the human genome, in its evolved form, is a stable and valuable possession that we must not tamper with. This view finds expression in the Universal Declaration on the Human Genome and Human Rights of the United Nations, which, as I mentioned in the introduction, declares the genome to be part of the "heritage of humanity."[37] Religiously, this view is echoed in the warnings voiced against playing God and in the belief that the human nature brought into being in Genesis is the ideal and normative form of human biological life. Leon Kass captures this when he says, "Man is the peak, both in possessing the *highest,* and also in possessing the *complete range* of, faculties of soul. . . . The story of the ascent of soul may already be complete."[38]

Octavia Butler disagrees. Although the "contradiction" humans exhibit, our combination of high intelligence and drive toward hierarchy, may have been adaptive in ancestral primate environments, it leads to catastrophe in the presence of advanced technology, be it guns or nuclear weapons. Hierarchy, in Butler's thinking, is synonymous with domination. It characterizes relations between men and women, between different human groups, and between human beings and nature.

In Butler's vision, the Oankali represent a positive biological and spiritual alternative to human beings. They are drawn to cooperation, collaboration, and union, not domination. They are incapable of and physically repelled by violence, even for self-defense (trade and violence are opposites). This quality makes them more "feminine" and is one of the reasons they are incomprehensible and even repulsive to many humans (although usually it is the more violent and racist men who feel this way). The Oankali cooperate and fuse with other peoples in genetic trade; they absorb information from all life and render life forms cooperative (as in their biologically grown starships). They are open to difference and to change. The Oankali do not believe in purity of race. At a point in *Adulthood Rites,* Lilith tries to explain this to Akin: "Human beings fear difference. Oankali crave difference. Humans persecute their different ones, yet they need them to give themselves definition and status. Oankali seek difference and collect it. They need it to keep themselves from stagnation and overspecialization. If you don't understand this, you will. You'll probably find both tendencies surfacing in your own behavior." As Lilith puts her hand on Akin's hair, she adds: "When you feel a conflict, try to go the Oankali way. Embrace difference."[39]

The moral and spiritual dichotomy between hierarchy/domination, on the one hand, and trade, on the other, between humans and Oankali, is deeply shaped by Butler's feminism and the American slave experience. The white slave-owners in *Kindred,* for example, exhibit many of the attitudes and behaviors evinced by the resister humans in her postnuclear world. They live within a series of hierarchies of domination (males over females; females over children; whites over blacks; some slaves over other slaves; humans over animals and nature). They are fearful of "difference," to which they respond with violence. They abhor genetic mixing and "uncleanness" (miscegenation), but they are perfectly willing to engage in sex across their own artificially drawn lines, as in rape, so long as it affirms and perpetuates their dominance. Finally, they make constant appeal to their own version of biblical religion to uphold the hierarchies of social, political, racial, and sexual power.

To some extent, the relationship between Oankali and humans also mirrors aspects of the slave experience. The Oankali closely supervise their human charges and unite with them sexually.[40] Yet in the entirety of the Lilith trilogy, the Oankali's authoritarianism is understandable, especially in view of human beings' dangerous, violent, and hierarchical tendencies. In the trilogy, it is not the Oankali but the human resisters who perpetuate the domineering attitudes and behaviors explored in Butler's writings about slavery.

A further theme of the trilogy is that trade is never without its risks. We see this in the Oankali's and ooloi's ambivalence about their powerful attraction to human beings. Although union with humans promises many benefits, including access to their intelligence and their ability to grow cells that normally lie dormant after birth (as in cancer), it also exposes the Oankali to the risks of emotional and social involvement with a violent species. At one point in *Dawn,* the ooloi Nikanj tells Lilith and Joseph, "You are horror and beauty in rare combination. In a very real way, you've captured us, and we can't escape."[41] Despite the risks, the Oankali never retreat from their willingness to reach out, bestow their gifts, and be changed by the genetic gifts of others. Their interstellar odyssey is a voyage of courage.

Most human beings do not possess such courage. Behind our hierarchies, propensity to dominate, rejection of difference, and quest for "purity" lies fear. Indeed, fear is our cardinal vice, the one that propels all the other deformations of the spirit to which we are heir. What Butler sees as our propensity to dominate is merely the flip side of our evolutionarily built-in fear of difference in others and fear of change in ourselves.

Octavia Butler's work does not answer many of the ethical, social, and religious questions that human genetic engineering raises. In a strict sense, she doesn't even address these questions, because genetic change in her fictional world is brought about not by weak and fallible human beings but by skilled and highly experienced interstellar genetic engineers. Nevertheless, if Butler's writings don't directly address these questions, more than any of the other imaginative texts on gene modification, they go to the heart of the issue. They offer a powerful spiritual alternative to some of our prevalent religious attitudes by holding out the prospect that our best future lies in openness to difference, change, and continuing genetic self-transformation. Dana in *Kindred* and Lilith in the trilogy, both the products of gene mixing, embody Butler's hopeful vision for our species.

Her vision makes us ask whether our negative reactions to the prospect of human genetic self-modification are based on reasonable ethical and religious concerns or irrational fears. Is the human genome so good that it should be regarded as unchangeable, or does our inherited biological nature possess flaws that should be corrected and traits that could be improved? Are the opponents of gene alteration fighting for human health and well-being, or is theirs an irrational fear of change? Is the urgent warning against playing God a wise reminder that we are fallible and limited creatures not capable of directing the course of our own evolution, or is it a way of perpetuating a genetic status quo that benefits winners over losers in nature's genetic lottery? Is the resistance to genetic change just a new expression of horror at racial "impurity" and the crossing of forbidden lines? Are our religiously driven genetic fears merely another instance of unexamined taboos? Octavia Butler's writings do not answer these questions, but they challenge the assumption that gene modification represents a spiritual and religious threat; instead, spiritual and religious arguments may represent an anxious defense of the genetic status quo.

The question stands: Should we play God with our genes? Answers, as we have seen, are sharply opposed, and the title of this chapter turns out to harbor at least three different perspectives on how humans and God should meet on the terrain of genetics.

In one perspective, human genetic engineering is seen as dangerous manipulation of a sacred reality, an impermissible *playing God* and *playing with* God's creation. Those who share this perspective see this technology as epitomizing the modern obsession with human autonomy and control. They view it as violating the integrity of nature and repudiating God's sovereignty over the world. Most who hold this view draw on religious sources, but they are sometimes joined by secular people committed to one or another progressive cause. On several occasions over the past twenty years, for example, Jeremy Rifkin, a leading opponent of all forms of gene modification, was able to persuade a diverse set of religious leaders and liberal intellectuals to sign petitions against germline gene therapy. Rifkin's coalition has included environmental activists and advocates of disability rights.[42]

The second perspective is much more optimistic. Its advocates see in human self-evolution the continuing story of increasing freedom from subjugation to nature and regard evolution itself as taking the next logical step when the evolved human creature becomes an agent of deliberate, designed evolutionary change. In the words of the Protestant theologian Ted Peters, "The human spirit is DNA's way of dreaming new futures."[43] This view is not irreligious. Peters and others emphasize our role as co-creators with God and see the process of human growth as involving not just spiritual growth but also biological self-transformation. They sometimes invoke biblical visions of a fully redeemed creation and a physically and spiritually new "Adam." They affirm that we are not born human but must become human. "This applies," says Peters, "to the whole human race as well as to individuals. We will become what we truly are only in the fulfillment of history, only at the arrival and fulfillment of the promised kingdom of God."[44] Those who hold this view interpret the title of this chapter differently. For them, human genetic engineering represents a laudable playing *with* God. We are God's partners in shaping the world around and within us.

Finally, there is the view of those, like Octavia Butler, who object to a religious framing of the question. They see human nature as imperfect, and they believe that we must take steps to change it. They criticize the religious outlook that sanctifies hierarchy, domination, and inequality, whether in social or in genetic forms. They urge us to *play* with God in the sense of having the courage to manipulate flaws in the world we live in.

The disagreements about intervening at the genetic level will grow in intensity in the years ahead. During the twenty-first century human gene modification is likely to move to the center of religious debates, possibly eclipsing the controversies about abortion, embryonic stem cell research, and cloning. Beginning with more widespread prenatal gene selection and moving on to germline therapies and enhancements, each new manipulation will precipitate a skirmish in the war between differing worldviews. The passions are strong, and the outcome of the debates probably depends on how well we implement the new technologies for choosing our genes. If we do so badly, gene modification will come under the shadow of the failed eugenics movement. If we implement it well, gene modification will become a routine and accepted part of our lives, joining anesthesia during childbirth, birth control, and in vitro fertilization on the list of reproductive technologies that religions once opposed.

CHAPTER 8

The Choices Ahead

In 1999, Alan and Louise Masterton lost their three-year-old daughter, Nicole, in a horrific accident in a Guy Fawkes Day bonfire at their home near Dundee, Scotland. Nicole was the baby of the family and had four older brothers. Soon after her death, the Mastertons sought to use fertility techniques to have another girl. They were not trying to replace Nicole. "We are mature and adult enough to know that we could never do that," said Alan Masterton. Rather, they were seeking to replace the "female presence" in the family. "We don't think there is anything wrong with boys," said Alan, "but it feels perfectly natural to us to have two genders in our family. We've had a female child; we know it does make a difference."[1]

The Human Fertilisation and Embryology Authority (HFEA), the official agency that licenses all infertility programs and procedures in Great Britain, rejected the Mastertons' request. The HFEA permits sex selection using preimplantation genetic diagnosis to avoid the birth of a child with a sex-linked disease like hemophilia or X-SCID. But the Mastertons' request for a girl to rebalance their family was a step too far. The HFEA chairwoman Ruth Deech said that the rules would not be relaxed for fear that it would open the door to using PGD to select for traits unrelated to disease. "The public do not like, and we do not like the idea of designer babies," said Deech.[2]

Following the denial of their request, the Mastertons became the focus of press attention. The House of Commons held hearings on the HFEA's policies. In their continuing effort to have a girl, the couple went to Italy, spending over thirty thousand pounds on three failed IVF treatments with sex-screened embryos. They joined the approximately twenty couples who leave Britain each week to have the sex selection procedure done abroad. Alan Masterton said that they would try again if they could afford it, and he asked, "Shouldn't we, as a loving couple and parents of a well-established family, decide what's better for our family?"[3]

The agency that denied the Mastertons' request, the HFEA, not only licenses infertility programs but oversees all public and private reproductive research and clinical services in Great Britain involving human embryos. The HFEA has been operating since the early 1990s, when it was set up following the advent of IVF. It is guided by British law (the Human Embryology and Fertilisation Act of 1990 and subsequent legal rulings) as well as public opinion. To assess public opinion, it sponsors "public consultations" in which the views of experts—physicians, researchers, bioethicists, and social scientists—are blended with those of ordinary citizens, which are gathered through polls and public meetings. Following a consultation in 2003, the HFEA reaffirmed its ban on sex selection, whether accomplished via PGD or by new preconception methods involving the sorting of sperm on the basis of whether they bear a male or female sex-determining chromosome.[4]

The HFEA's position on sex selection, which the Mastertons ran up against, stems from many concerns. One is the global alarm about the use—or misuse—of reproductive technology for sex selection in developing countries like India and China. In these nations, cultural and religious values as well as economic forces have worked to create widespread preference for sons, with serious imbalances in sex ratios as a result. In some poor regions of India, the ratio of males to females is 130 to 100 (the normal ratio is 105 to 100). Some of the preference for sons arises from the increasing emphasis on dowries, even among less affluent families, which makes daughters an economic liability. In China, the "one child per family" policy has contributed to the problem because most parents want at least one boy. Both countries have passed laws against the use of fetal ultrasound technology for sex determination and selective abortion, but the laws are hard to enforce because parents can pretend that they want to use the technology for the prenatal diagnosis of disease. The result has been an increase in the already-large deficit in female births caused by the neglect or infanticide of female babies. In 1990 the Nobel Prize–winning economist Amartya Sen estimated that there were about a hundred million fewer living women than would have been produced without sex selection. Although others have offered a somewhat lower number, there is widespread concern that the imbalance in the sex ratio could mean that before long, vast numbers of males will be unable to find women with whom to have children.[5]

But Britain is not India. Although it has immigrant Asian subcommunities, some of whose members bring their sex preference for sons with them from the home country, most of the opposition to sex selection draws on other sources. One is the fear of a slippery slope. As Ruth Deech suggested, if we permit genetic selection for sex, won't this lead to testing for other nondisease traits? The same worry was raised several years ago in connection with the requests for savior children that I mentioned in chapter 2. These children are conceived by IVF and selected as embryos via PGD to match the HLA or immune system profile of an existing child suffering from a disease that can be treated with a matching bone-marrow transplant. At first, the HFEA adamantly refused to permit this on the grounds that such requests required the genetic testing and selection of embryos for something other than the embryo's risk of disease. But after a wave of requests and trips abroad by many desperate couples desiring the service, the HFEA relented. Today, PGD to find a matching sibling donor is routine in Great Britain, as it is in the United States. Prenatal testing for a nondisease trait is already widely practiced, then. Of course, testing for savior children is at least related to the treatment of disease—in this case, the sibling's disease. But the HFEA's concession shows that gene selection for traits can be justified for compelling reasons.[6] The question is why family balancing, as in the Mastertons' case, where the parents wish to have at least one child of each sex, is not also a compelling reason. Since we are already on the slippery slope, why draw the line to exclude sex selection?

There are other concerns. Although polls show that parents in developed countries show little preference for either boys or girls, there is a worry that if parents are free to select for sex, any preference for boys, however small, could have negative effects.[7] Studies show that being firstborn provides an advantage in terms of parental attention and later success in life (although later-born children can also sometimes profit from additional parental attention).[8] Would permitting sex selection further increase males' advantages over females in society? To address this concern and minimize the problem, programs that offer sex selection can require parents requesting the service to have had at least one child of the opposite sex. This places the emphasis on family balancing rather than meeting parents' sex preferences.

Some believe, further, that sex selection, even for family balancing, stems from—and fosters—inherently sexist attitudes. Why do I want to have a child of a specific sex unless I think that sex is important? And doesn't this attitude usually correspond to prevailing cultural stereotypes, especially those that privilege boys and discriminate against girls? A number of bioethicists have advanced this argument, but no one has put it more forcefully than the philosopher Michael Bayles in his book *Reproductive Ethics*.[9] Bayles offers the case of a hypothetical individual, Jeremiah, who has two girls and seeks help in having a boy. "But why would two daughters and one son be preferable to three daughters?" Bayles asks. Jeremiah might respond by saying that he would like a son so that he could go fishing or play ball with him. But that preference, says Bayles, rests on a sexist assumption:

> As the father of two daughters, I have fished and played ball with them, watched my daughter play on a ball team, and gone camping and hiking with them, as well as cooked, cleaned house, done laundry, and engaged in various other so-called women's activities with them. There may be some activities that are strongly sex-related in that members of one sex are generally better at them than members of the other sex. For example, perhaps most women have a greater aptitude for ballet than men. Recognition of such differences in role aptitude is not sexist, but the assumption that no members of the other sex can perform the same roles well or that one set of such roles is preferable to the other is sexist. Thus, a desire to have a male (female) child because of a preference for one set of "sex-linked" roles is sexist. Nor can one argue that variety is desired. Were children allowed to develop freely their own interests and talents, children of the same sex would probably exhibit as much diversity as children of opposite sexes. . . . In sum, sex preselection on the basis of intrinsic sex preference is always wrong.[10]

Bayles's argument is not convincing. There are probably people like Jeremiah whose requests for help in selecting their child proceed from sexist assumptions. But there are also many others, like the Mastertons, who merely want a female (or male) presence in their family. Some of our preferences for one sex or the other are based on harmful stereotypes—like the view that only boys can play baseball—but others are the unavoidable result of differences in biology. This includes differences in bodily form, hair growth, voice development, and sexual maturation. Unless we think that we should take steps to utterly erase sex differences—for example, by dressing boys and girls alike—those differences will inevitably enter into our culture and shape parental aspirations. Is it really sexist for a woman to look forward to taking her son for his first haircut—or to dream about the day when she can assist her daughter during her first pregnancy and birth? Is it sexist for a father to take pride in giving his son his first razor or to anticipate the day when he can stand beside his daughter at her wedding? If we think not, then we can begin to see that the arguments against family balancing are strained.

I have gone into the Mastertons' case at length because I think it is a lens through which we can begin to perceive the value—and the problems—of having the government regulate familial genetic decision making. The case raises the question of how we can best ensure that new developments in reproduction and genetic medicine are used humanely and do not escape our control. In some ways, Britain's HFEA seems to be an ideal solution. It draws its positions from democratic decision making. It has the power to oversee all applications of these technologies within the nation, whether public or private, and it can intervene to shut down procedures deemed to be unsafe or unethical. This contrasts sharply with the much more chaotic situation in the United States. Here, on the one hand, federal regulations governing aspects of embryo research are far more restrictive than in Britain. On the other hand, privately supported research and clinical care are largely unregulated. Infertility clinics remain free to develop and introduce whatever services parents are willing to pay for. The world of American reproductive medicine has been described as a "cowboy culture."[11]

And yet, as the Mastertons' difficulties show, the British model is not perfect either. The HFEA intervened to prohibit reproductive genetic procedures only to retreat when its positions were challenged by events. British citizens have frequently been forced to travel abroad to obtain services unavailable at home. In some cases, couples have initiated lawsuits based on the European Community's human rights standards to overturn HFEA regulations. In the United States, in contrast, new procedures have been introduced with little fuss—for example, there was never much controversy in the United States about savior child testing. And even though many people, including most genetic counselors, oppose sex selection for nondisease reasons, opposition has not crystallized in a refusal by all infertility clinics to provide the service. One recent study of 415 infertility clinics in the United States showed that 42 percent of them were offering sex selection for non-medical reasons.[12] This provides options for American—and for British—couples like the Mastertons. A good example of the responsible use of this freedom is Baylor School of Medicine's infertility program, which offers sex selection for family balancing in the context of a research study. The program opens its doors to people who have had at least one child of either sex and wish to have another child of the opposite sex. The investigators are keeping careful track of the parents' backgrounds and the reasons they offer for wanting the service. Baylor's aim, to understand why parents are seeking this service, is a way of beginning to assess the impacts of such choices on family and society.[13] Here, instead of the stern no that the HFEA began with, we find an information-gathering yes.

The strengths and weaknesses of the British and the American models are worth keeping in mind as we think of the choices ahead. Many people, concerned about the possible abuses and risks of repro-genetic technology, have called for stringent regulation. The activist Jeremy Rifkin angers many people in the biotechnology sector with his frequent diatribes against all genetic engineering, but he struck a popular chord when he called for a moratorium on all human gene therapy research until such time as the National Institutes of Health establishes a "Human Eugenics Advisory Committee" to evaluate its implications.[14] Francis Fukuyama has become one of the most outspoken defenders of strict, centralized regulation. In his book *Our Posthuman Future,* Fukuyama asks, "What should we do in response to biotechnology that in the future will mix great potential benefits with threats that are either physical and overt or spiritual and subtle?" The answer, he says, "is obvious. *We should use the power of the state to regulate it.* And if this proves to be beyond the power of any individual nation-state, it needs to be regulated on an international basis. We need to start thinking concretely now about how to build institutions that can discriminate between good and bad uses of biotechnology, and effectively enforce these rules both nationally and internationally."[15] In 2006, Fukuyama and a colleague, Franco Furger, issued a lengthy report calling for the creation of an "independent" executive branch commission that would regulate all the new reproductive technologies.[16]

But the Mastertons' case is a reminder that what seems obvious is not always right. Here, powerful parental wishes ran up against an unresponsive regulatory authority. In the end, the Mastertons and other couples like them traveled abroad to escape rules that made no sense to them. Despite Fukuyama's optimism about an international solution, that could raise the problem to a higher level. When parents do not agree with the authorities' rulings and when they strongly believe that a new reproductive service will benefit them or their child, they will do what they must to evade the restrictions: travel to a jurisdiction that refuses to comply with the international rules, or go underground and put themselves in the hands of renegade practitioners. Legal bans could also create a black market in new reprogenetic technologies, increasing the likelihood that only affluent people will have access to them.[17]

The philosopher Mary Anne Warren points to the limits of enforcement in matters like sex selection. "Because the forbidden actions are not in themselves overtly harmful, many people will continue to believe—with a great deal of reason and justice—that they have a right to perform those actions; and many will continue to do so regardless of the law." This reasoning leads to what she calls the "paradoxes of unenforceable prohibitions."[18] Regulations that fail to elicit widespread support not only are ineffective but also, by inviting lawbreaking, tend to aggravate the very ills they are designed to prevent. The global nature of technology makes enforcement particularly difficult.[19]

The regulatory solution has other serious problems. In free societies the presumption is that the burden of proof lies with those who would restrict others' liberty. But in a regulatory environment, restraint easily prevails. Fears and apprehensions catch the attention of regulators and dominate the conversation. The view that sex selection is inherently sexist has had so much influence in the debates for precisely that reason. The mere hint of sexism has been enough to dissuade people from taking requests like the Mastertons' seriously.

If these problems strain the work of Britain's HFEA, they are even more troublesome in the United States. With large evangelical Protestant and Roman Catholic populations, the United States is far more religiously divided than Britain is on issues related to human sexuality and reproduction. It is nearly impossible to establish here the kind of independent commission that scholars like Furger and Fukuyama dream of. The reality is that from the appointment process on, a determined minority can easily seize the levers of political power and block the development of anything they oppose. I learned this during the mid-1990s, when I served on the Human Embryo Research Panel of the National Institutes of Health. Our challenge, after a two-decade-long de facto moratorium on any research related to the human embryo during the Ronald Reagan and George H. W. Bush presidencies, was to come up with rules to guide new research initiatives under consideration by the Clinton administration. To my dismay, I witnessed the power of determined opposition groups like the Roman Catholic Bishops' Secretariat for Pro-Life Activities and various Protestant evangelical groups to block initiatives in this area. In 1994 our panel recommended federal support for stem cell research, but a newly elected conservative Congress, influenced by a well-organized minority espousing conservative religious opinion, quickly rejected our recommendations. Four years later, Jamie Thompson at the University of Wisconsin, working in a privately funded laboratory that he had created to keep his research separate from the university's publicly funded facilities, isolated the first human embryonic stem cell lines, and the debate about federal support for stem cell research began in earnest. There is still little federal support for human embryonic stem cell research, even though opinion polls show that a large majority of Americans favor it. Whatever research is now going on occurs in the private setting or, most recently, in several states, like California, where voters have chosen to fund it. Significantly, Jamie Thompson's breakthrough research could never have happened had the federal ban on stem cell research extended to the private sector.

This experience shows that an HFEA-like regulatory body in the United States could easily become a tool for the most determined opponents of reprogenetic research. Since many people on the conservative side of the spectrum oppose any manipulations of human embryos, it is also likely that a national regulatory body would prohibit many current clinical practices, such as the use of multiple embryos in IVF and most forms of preimplantation genetic diagnosis. In 2004, Italian legislators, as a result of strong ecclesiastical pressure, passed a law prohibiting couples from producing more than three embryos, all of which must be transferred to the womb. This has significantly reduced the success of IVF in that country and forced couples to go abroad seeking better services. Because genetic interventions tend to unite extremist opponents on both the right and the left sides of cultural debates, a federal regulatory body whose powers reached beyond the area of federally funded research would probably become a major obstacle to advances in genetic or reproductive medicine and would repeatedly impose unnecessary restrictions on parental choice. Earlier, I disputed the concern that giving parents the freedom to shape their child's genetic inheritance would necessarily lead to a resurgence of eugenics. Ironically, those who would restrict parental freedom in this area come closest to thinking in a eugenic way. As Julian Savulescu reminds us, "The lesson we learned from eugenics is that society should be loath to interfere (directly or indirectly) in reproductive decision making."[20]

What is the alternative? Clearly, not everything that parents might want is right. We need both norms and oversight for the powerful new reprogenetic innovations. But how can these be put in place without creating a super-HFEA, which may heed only the loudest or most politically favored voices? How can we preserve the freedom of parents to make responsible decisions in this area without subjecting them to uninformed majority (or determined minority) tyranny?

The answer, I believe, lies in creating a system that is pluralistic and diverse, one that provides overlapping methods of oversight and restraint but that is also open-textured enough to make responsible innovation possible. Instead of one all-powerful national HFEA or Human Eugenics Advisory Committee, there should be a mixture of federal oversight, state regulation, common law protections, and significant reliance on noncoercive instrumentalities, like information gathering and education, to empower parents to make responsible decisions.

Elements of this system are already in place.[21] The Recombinant DNA Advisory Committee, mentioned in earlier chapters, was established in 1974 by the National Institutes of Health in response to the first wave of concerns about genetic engineering in bacteria and plants. The RAC was conceived as a high-level body that could provide oversight and approvals for all research involving the new genetic technologies, from the creation of new bacterial organisms in the laboratory to human gene therapy. Although the RAC's authority extends only to federally funded research, its opinions and rulings have great force in the private sector as well because companies and private-sector researchers know that adherence to RAC guidelines can help win public acceptance of their work.[22] In the mid-1990s the NIH director, Harold Varmus, limited the RAC's scope to novel research protocols and removed its authority to block research, handing that over to the Food and Drug Administration. The RAC continues to operate as a high-level advisory committee that reviews the policy and ethical dimensions of gene therapy research. Its members include experts drawn from the scientific and medical disciplines, as well ethicists and members of patient and other lay communities. Its meetings are open to the public, and its opinions, though lacking regulatory force, are widely heeded.

The FDA complements the work of the RAC. Its authority extends to all new drugs, devices, and genetically engineered agents, including those used in somatic cell gene therapy, whether in federally funded or private-sector research and therapy. In an advisory in 1993 the FDA stated that its existing statutory authorities, "although enacted prior to the advent of . . . gene therapy, are sufficiently broad in scope to encompass these new products and require that areas such as quality control, safety, potency, and efficacy be thoroughly addressed prior to marketing."[23] Unlike the RAC, the FDA operates under rules of confidentiality because it must deal with information of value to private companies. Although the FDA has not claimed authority over germline genetic research or therapeutic agents used in it, in

2001 the agency intervened when researchers at St. Barnabas Hospital in New Jersey announced that they had been using a technique known as ooplasm transfer to help several women with impaired fertility bear children. This involved inserting material from the cytoplasm of the egg of a fertile woman into the egg of a woman experiencing infertility. The hope was that the healthier cytoplasm would jump-start the infertile woman's eggs.[24] Since egg cytoplasm contains the cell's mitochondria, tiny energy-producing bodies with a small amount of DNA, the changes in the egg would be transmitted to all the cells of the resulting child and would be inheritable by its descendants. The FDA was not pleased when the researchers proudly described their work to the press as the "first case of human germline modification."[25] Following the St. Barnabas announcement, the FDA notified all U.S. fertility clinics known to be offering the procedure that further ooplasm transfer protocols could not proceed without the agency's approval.[26]

The FDA's oversight is not complete. Because it focuses largely on the safety and efficacy of therapeutic agents, it lacks the RAC's authority to consider the ethical and social implications of introducing them into widespread use. Furthermore, once a drug or biologic agent is authorized for medical use, the FDA cannot prohibit qualified medical practitioners from applying it for other, "off-label" purposes. In its current functioning, the FDA has also been criticized for its post-approval, post-marketing follow-through, which relies almost entirely on manufacturers and physicians to report adverse events resulting from the use of FDA-approved drugs and therapeutics.[27] Together, the RAC and FDA provide a web of regulatory protection against extreme abuses. Preserving this relative complementarity of approaches may be better than uniting them into the single regulatory superauthority desired by people like Rifkin and Fukuyama.

Other sources of control are built into the American regulatory system. Local institutional review boards (IRBs) at medical centers and universities doing federally sponsored research must scrutinize research protocols to make sure their human subjects are well protected. This includes careful oversight of the informed consent procedures. Since the IRB system went into effect, critics have voiced concerns about placing so much authority in the hands of local review boards composed of scientists, clinicians, bioethicists, and community representatives. But although local review has some problems, it has the advantage of keeping reviewers close to the real conditions affecting research in their institution.[28] In addition, members' proximity to the research community makes them less vulnerable to national political currents. This system, though imperfect, has served us well.

Apart from the FDA, none of our major national regulatory agencies has the authority to reach into purely private-sector research or therapy, such as that going on at fee-for-service infertility programs. These programs, which usually are eager to expand their services, have already begun to offer PGD for genetic diseases, infertility problems, and sex selection. They are the most likely location of future gene modification attempts. If their efforts do not involve a biologic agent subject to FDA approval, they are largely free of regulatory oversight. This seems to be a major hole in our protective network. Some professional bodies, like the American Society of Reproductive Medicine and the American Medical Association, have tried to fill it by promoting professional standards of conduct. But their guidelines are subject to criticism for being the product of very groups that need oversight and for usually lacking ways to punish practitioners who violate them.[29] There is also a tendency where controversial new technologies are concerned for professional bodies to take very restrictive positions in order to protect the profession's reputation. In either case, parents and families are not well served.[30]

By confining our thinking to regulation, we miss a major resource for the legal and ethical control of reprogenetic technologies: common law. Our legal system has provisions for litigation that can punish harmful research and clinical activities.

Two kinds of lawsuits have already cropped up in response to the negligent or reckless practice of reproductive medicine. The first deals with "wrongful birth." It is usually initiated by parents who believe that a medical practitioner has wronged them by causing them to have a child they did not want. Typical cases in this area involve faulty contraceptive advice or mistaken genetic diagnoses, as when a counseling team fails to inform parents that they are carriers of a known genetic disease like Tay-Sachs. In such instances, parents are denied the opportunity to consider avoiding or terminating a pregnancy and wind up with a child whose care is emotionally burdensome and expensive. Wrongful-birth lawsuits are common, and when medical professionals depart from the recognized standard of care, parents often win.

The second kind of litigation, known as a "wrongful-life" lawsuit, has been much less successful. Here, it is the child who, having been born with a birth defect or genetic disease, initiates a suit against the practitioner. In the United States these cases have usually arisen because such legal technicalities as statutes of limitation prevent the parents but not the child from suing. In a wrongful-life suit, the child claims that life is so unsatisfactory that he would have been better off if he had never been conceived or born. The child holds the practitioner responsible for failing to provide his parents with the information needed to prevent his conception or birth.

Wrongful-life suits have been won in France and Israel, but, with a few uncertain exceptions, British and American judges have resisted them.[31] There are two concerns. One involves the claim that being born has harmed the child. The concept of harm in injury law usually relies on the idea that the injured party or plaintiff has been made worse off than he or she would have been if the defendant had not acted wrongly. But U.S. and British judges have been reluctant to rule that a child born with a serious impairment is worse off than the child would have been if the child had never existed. In one early wrongful-life case, *Berman v. Allan,* the court rejected a wrongful-life claim on the grounds that "life—whether experienced with or without a major physical handicap—is more precious than non-life."[32]

A second problem that bothers some courts is that wrongful-life suits could open the way for children to sue their own parents. If I can sue a doctor for not telling my mother about a genetic test that would have allowed her to make a decision not to have me, why can't I also sue my mother if she willfully ignores the doctor's advice and brings me into being with a serious impairment? But if children can sue their parents for the hardship of their life, where will it end? Won't suits by angry teenagers flood the courts?

In fact, neither of these problems is as troublesome as critics contend. Wrongful-life lawsuits, I predict, will become a useful resource in the future for controlling the irresponsible use of reprogenetic technologies by both medical practitioners and parents.

The first problem, that it is impossible to compare impaired life with nonexistence, arises only because of a confusion in thinking. The mistake is to think of nonexistence—in the sense of never having been conceived or come into existence—as the same thing as dying. We all struggle to avoid dying, but not being conceived is not a harm, nor is being conceived a benefit. Before conception at least, there literally is no one who is the object of our moral concern, no one to injure or help.[33] If we are tempted to think otherwise, and mistakenly slip into the view that we benefit a child by conceiving it, what are we to say about the billions of children never born because potential parents used birth control or just avoided sex? A classic *Peanuts* cartoon justifiably pokes fun at this confusion. After Lucy loudly questions whether it is right to bring children into "this uncertain world," she is challenged by Linus, who asks, "What are you gonna do with all those babies who are lined up waiting to be born?" Lucy is dumbstruck, but the humor derives from the silliness of Linus's question.

Once we see that coming into existence is not a benefit, then the balancing of the benefit of life against an impairment that some judges have worried about is not a concern. All we need to ask, once a decision is made to bring a child into being, is whether we believe that parents and the professionals who help them should be required to do their best to avoid serious injury to that child. Do I wrongfully harm a child if I knowingly (or negligently) expose the child to congenital damage? For example, does an obstetrician act wrongly who fails to inform a woman that she should take a folic acid supplement before conceiving in order to protect her child from serious spinal defects? Does the woman act wrongly if she willfully ignores the doctor's advice? The answer to both these questions is, I believe, yes. Most of us think that, other things being equal, parents and medical professionals should try to give a child a healthy start in life. The alternative, careless reproductive behavior, causes too much harm. No one has to be born. A child is not wronged if the child never comes into being. But if a child is born, why allow negligence or thoughtlessness to cause injury? In concluding that we should avoid needless harm to born children, we are not weighing children's impaired lives against children's nonexistence; we are weighing the impaired lives they have against the healthy lives they could have had if the parents or doctors had acted responsibly.

Parents can still have valid competing interests of their own. If there is no other way that two people can have a child without risking some injury to it, as when both carry a recessive genetic disorder, we normally believe that they have a right to try to have the child, even if they risk its health. The right to be a parent is a major factor in our thinking. But because parents must have a good reason like this for inflicting injury, there is also a moral obligation to strive to reduce the chances of harm. The same is true when a medical professional is involved. A doctor who has acted negligently should be held financially responsible for causing a child to be born with a significantly reduced health status compared with what the child would have had but for the doctor's malpractice.

The second concern is that wrongful-life suits could lead to children suing their own parents. But in terms of reprogenetic medicine, this is less a problem than it is an advantage. Admittedly, the thought of parents being sued by their child is distasteful. But if the parents have acted without regard for the child's health, why shouldn't they be called to account? This possibility has already arisen in connection with cloning. Although the technology is far from proven safe, and any child born of it is likely to suffer grievous birth defects, attention-getting physicians like Panayiotis Zavos and Severino Antinori have offered the service, and religious cults like the Raelians have recruited their members as volunteer parents for cloned children. Although passing laws against these reprogenetic opportunists would be appropriate, litigation also has a place. If the doctor misrepresents the risks and fails to deliver the promised healthy child, parents could undertake a malpractice lawsuit on the child's behalf. If the parents themselves are fully and knowingly complicit with these ill-advised attempts, why shouldn't they (and their religious cult leaders) be held liable?

Major legal complexities abound, of course. Courts will have to discover ways of keeping frivolous wrongful-life suits in check, probably by placing a heavy burden of proof on litigants and by giving parents the benefit of the doubt in child-parent lawsuits. Where parents are held liable, courts will have to identify ways of awarding damages that do not penalize the child. But courts in the jurisdictions where these suits have been allowed have already begun to chart this ground. Despite the challenges, wrongful-life litigation is an alternative to cumbersome legal prohibitions and to regulations that lag behind realities or are waylaid by special interest groups. Opportunistic politicians have repeatedly demonized trial lawyers, but the reality is that medical malpractice litigation has dramatically improved public safety. For example, following a wave of lawsuits in the 1970s and 1980s, anesthesiologists began safety improvement programs that reduced the risk of death from anesthesia from 1 in 5,000 to about 1 in 250,000 today.[34] I believe that wrongful-birth and wrongful-life litigation, through both punitive and deterrent effects, will be some of the most valuable tools for controlling harmful conduct by parents and practitioners in the area of reprogenetic medicine. In her 1982 presidential address to the American Society of Human Genetics, the geneticist-lawyer Margery Shaw put the matter succinctly when she said, "Future generations will be the beneficiaries of our increasing predictive powers and therapeutic tinkering. Parenthood may become a privilege to be cherished rather than a right to be exercised even when a child is harmed."[35]

Wrongful-birth and wrongful-life litigation, like all injury law, draws on a wealth of moral judgments. When lawyers, judges, or juries conclude that someone has engaged in "unreasonable behavior" or has "wrongfully harmed" another person, they appeal to our shared moral convictions. This tells us that we need a basic set of moral rules or guidelines to help us develop laws and regulations and to decide how those laws and regulations should apply to specific circumstances. If our basic norms are mistaken or incomplete, nothing that we build on them will be right. Bioethicists like Allen Buchanan, Dan Brock, Norman Daniels, and Daniel Wikler have offered a series of considerations to aid us, as has the British-Australian bioethicist Julian Savulescu. Drawing on their work and others', I would like to offer four ethical guidelines for human genetic self-modification. More guidelines will probably be needed in the future. But for now think of these as stars that can help guide our course as we move into this uncharted area of moral choice.

First Guideline: Genetic interventions should always be aimed at what is reasonably in the child's best interests. In the novel *Never Let Me Go* (2005), the Japanese-born British author Kazuo Ishiguro traces the maturation of Kathy H. from her student years at Hailsham, an elite British boarding school, to her adult years as a caregiver for Hailsham graduates and others nearing death. As the novel unfolds, we learn that Hailsham students are clones bred to be organ donors. The novel ends on a bittersweet note, with Kathy caring for her lover Tommy, still a young man, as he enters the final, lethal phase of donation.

The cruel and exploitive values depicted in the world of Ishiguro's novel (and also portrayed in the 2005 film *The Island*) are directly opposed to those that should govern genetic interventions. As the Christian ethicist Sondra Wheeler puts it, "Any proposed intervention must pass the tests of being undertaken directly and primarily for the sake of the child rather than the parents or other parties."[36] Genetically modified children—babies by design—are ends in themselves, and whatever is done to select or alter their genomes must be done in ways that can reasonably be judged to be for the child's own good.

The word "reasonably" here is important. It both constrains parents and empowers them. It constrains them because it implies a larger standard of judgment than mere parental wishes or beliefs. It is not enough for parents to *think* that something is in their child's best interests; their judgment must accord with that of most other informed members of society. We routinely apply this reasonability test to parental choices when we override harmful parental decisions. For example, courts have consistently ordered children of Jehovah's Witnesses to undergo lifesaving blood transfusions, even when the parents oppose it. Jehovah's Witnesses parents almost always sincerely believe that they have the best (spiritual) interests of their child at heart. But as a society we have concluded that forcing an unconsenting minor to sacrifice his life in the name of the parents' beliefs is unreasonable. Surviving to make his own spiritual choices in adulthood better protects the child's interests. Similarly, a father may believe that he is acting in a child's best interests when he seeks to modify the child's genes to give him enough height to vicariously fulfill the father's dreams of a professional basketball career. But this modification may be associated with a greatly increased risk of heart disease or stroke. As such, the request does not clearly serve the child's best interests, and this first moral guideline tells us that medical professionals should not honor it.

If the word "reasonably" limits parents by invoking a standard of judgment, it empowers them by saying that their decisions should prevail so long as they are not clearly unreasonable. In our treatment of parenting in chapter 5, we saw that parents' dreams and aspirations have a legitimate role to play in shaping a child's life. Parents are both guardians and gardeners. As guardians, they cannot inflict serious harm on the child. But neither are they obligated, as some believe, to keep the child's future absolutely open. In their gardening role parents have the right, and even the obligation, to impart their values and ambitions to their child and, within reason, to try to equip the child with the abilities needed to uphold those values and pursue those ambitions. Equipping the child can extend to gene modifications that shape the child's values and skills. An athletically inclined family, for example, might seek to use genetics to offer added abilities to their children. If Tiger Woods can use LASIK surgery to acquire 20/15 vision, freeing him from the need for glasses or contact lenses in a demanding sport, why can't parents interested in golf or skiing choose to confer a similar visual advantage on their child by reprogenetic means?[37]

This example is important. It tells us that parents can best accomplish both their guardian and their gardener roles by opting for enhancements that are useful in any of the possible lives that their child may choose to live. Dan Brock and Norman Daniels call these "general" or "all-purpose" means. They describe them as "capabilities that are broadly valuable across a wide array of life plans and opportunities typically pursued in a society like our own."[38] Interventions of this sort are so benign and so useful that Brock and Daniels even see a role for the government in encouraging or requiring them. As a precedent, they point to widely accepted water fluoridation programs that provide children with greater-than-normal resistance to tooth decay.

One test of whether parents' choices are in the child's best interests is whether the child is likely to accept the parents' decisions when he or she grows up. This is a useful guide to choice as a rule of thumb, but it is not perfect. The citizens in Huxley's *Brave New World* fully accept their programmed role in the World State's totalitarian order, but they have been manipulated to make them do so. Similarly, we can imagine a religious sect whose members are opposed to the values of the modern world and who retreat to a simple agrarian existence. The sect might be tempted to use genetics to reduce their children's cognitive abilities in order to suit them to the sect's limited environment and to make them obedient to their elders. Since these choices irreversibly narrow the children's range of options and make them more vulnerable to harm, they are not in the children's best interests. Even if some of the children later consent to what has been done, this does not make it right. In other words, a child's likely consent is a rough-and-ready first test, but it should always be measured against the broader standard of what the larger community regards as reasonably being in the child's best interests.

Second Guideline: Genetic interventions should be almost as safe as natural reproduction. In its earliest phases, germline gene transfer re-

search for therapy or enhancement will involve risks for any child produced with these technologies. *These risks are our leading concern about
such interventions.* But what level of risk is morally acceptable? This
question is more easily answered for germline gene modifications to
prevent disease, since the dangers of a disease like cystic fibrosis or
sickle cell anemia justify the risk of trying to forestall it. Pure gene enhancements complicate the picture. Here the parent is trying to provide added benefit for what would otherwise be a normal child. Is it
allowable to expose a child-to-be to risks just to give it added capabilities, such as sharper vision, a more attractive physical appearance, or
enhanced cognitive abilities? If so, how much risk is allowable?

Some would insist on a no-risk standard. During the early 1970s
conservative bioethicists like Leon Kass and Paul Ramsey argued
against developing IVF because of possible risks to the children that
could be born as a result. Kass and Ramsey observed that the initial attempts at IVF would be a leap into the dark. Even the most extensive
preliminary animal research could not ensure that human IVF would
not damage a child, possibly in subtle ways that would not be detected
until later in life. These risks, they said, were being imposed merely to
satisfy the parents' desire for a biologically related child. But in research ethics it is generally regarded as wrong to expose children to
more than minimal risks for others' benefit. This led Kass and Ramsey
to the conclusion that all such reproductive research amounted to
"unethical experiments on the unborn."[39]

Some might reply to these arguments by saying that if the child
would not be conceived or born in the absence of the new techniques—for example, if parents could or would not otherwise have the
child—then the benefits to the child of coming into being outweigh the
risks. The legal scholar and bioethicist John Robertson has frequently
defended risky new reproductive procedures in this way.[40] But as I
have said, I believe this argument is a dead end. Being brought into be

ing does not benefit the one brought into being. Linus aside, it does not assist our thinking to imagine babies as lined up in heaven waiting to be born.

But the obligation to minimize harm to a child while bringing it into being does not mean, as Ramsey and Kass believe, that researchers and parents should be held to a no-risk standard. Even natural reproduction has its risks, including the physiological risks of gestation and birth as well as genetic risks associated with fertilization and early embryonic development. It has been determined, for example, that between generations about three new single nucleotide mutations occur in each individual, mutations that can potentially cause an entire gene to malfunction.[41] In addition, chromosomal translocations and deletions affecting millions of DNA letters sometimes occur.[42] The physician-essayist Lewis Thomas, commenting on the role of genetic mutations in driving the evolutionary process, once remarked, "The capacity to blunder slightly is the real marvel of DNA. Without this special attribute, we would still be anaerobic bacteria and there would be no music."[43] Unfortunately, the vast majority of nature's blunders are harmful, and children born with them can suffer serious illnesses. Thomas's remark reminds us that it is wrong to think of conception and birth as harmless processes that only our reprogenetic interventions disrupt. The risks are already there.

Some might object that although reproduction has its risks, novel enhancement procedures add another layer of risk that we are not justified in imposing. But this misses the larger point. We allow parents to try to have children despite the risks in every pregnancy because we believe that parents have a right to try to have children of their own. This right guided our thinking back in the 1970s and 1980s when the first attempts at IVF were made. Despite the appeals of people like Ramsey and Kass for a no-risk standard, we recognized that a measure of added risk was justified if it helped infertile couples have children. This acceptance of risk still holds today, when even well-established infertility techniques carry such added risks as a higher incidence of multiple and premature births.

The same logic applies to gene modifications. Parents' wishes are an important part of our moral equation. They have weight and should be respected so long as the child is not likely to be seriously harmed. It is a measure of how much we respect parental reproductive liberty that society does not prevent couples at risk of transmitting a severe genetic disease like cystic fibrosis from trying to have a child of their own.[44] In addition, where enhancement is concerned, we should factor into our thinking the prospect of added benefit for the child. If rational adults can invite some risks in undergoing cosmetic plastic surgery or a laser eye procedure, parents can also accept some added risk for their future child to give it these benefits.

How much risk? As a rule of thumb, enhancement research and clinical implementation should proceed only when the risk levels are no greater than they were in development of IVF—that is, when they represent only a slight increment over the normal risks of pregnancy and birth. To ensure this requires solid preliminary research using animal models and the confidence that the move to human beings involves no major unknowns.

To achieve these risk levels, several further guidelines come to mind. First, germline genetic modifications should normally be a last rather than a first resort. Whenever we can efficiently accomplish a result using pharmaceuticals, we should do so rather than attempting more fundamental gene modifications. In the future new drugs will be available for enhancements like improved disease resistance, mood, and even cognitive abilities. We are already treating children with human growth hormone and certain mood- or behavior-altering medications. Using drugs and gene products rather than changing the genes themselves has the advantage of increased control, and it permits children or future adults to participate in the decisions affecting them. The same reasoning should lead us to privilege somatic cell gene interventions over germline ones. Even though these treatments may have to be repeated often during a person's lifetime as the altered body cells wear out and again in each unaltered subsequent generation, the lack of permanency has advantages. Nevertheless, certain enhancements, such as cognitive improvements, may have to start in the earliest phases of embryo development. As science increases our understanding, the acceptability of germline enhancements grows.

Second, the initial interventions, whether for germline changes or for enhancements, should be also reversible, using the kinds of technologies developed by researchers like Mario Capecchi. Recipients of altered genes should be able to take a drug that erases the modification from all the cells in their body and, in the case of enhancements, replaces it with the prior, normal sequence. Other newly developed techniques could also be used to stop the functioning of altered gene sequences inside cells.[45] One of these, RNA interference (RNAi), aims at blocking the functioning of the messenger RNA that carries instructions from the DNA in the genome to the cell's machinery for making proteins. Using this technique, clinicians could selectively shut down the production of a protein made by the cell's modified DNA without affecting other proteins made by the same DNA. Approaches like these would permit us to minimize the damage done by changes that prove harmful, and they would also give enhanced children the power to reject the decisions their parents made.

Finally, until we gain solid confidence in the long-term safety of any gene intervention, there should be ongoing follow-up of any children produced in this way—and of their children as well.[46] I mentioned the problems that the FDA currently has with postmarketing surveillance of pharmaceutical use. If the agency is going to provide oversight for genetic procedures, it and its partners in the provision of reprogenetic medicine (biotech companies, health-care providers, professional medical organizations) are going to have to assume a corresponding responsibility to monitor the recipients of procedures throughout their lives. And until the safety of specific germline procedures is established, follow-up should also continue into subsequent generations. Such follow-up might be seen as intrusive by those future people who, after all, had no part in the decisions that led to the need for it. Nevertheless, this degree of continuing medical surveillance is warranted in view of the overriding public interest in identifying and forestalling serious health risks.

Third Guideline: We should avoid and discourage interventions that confer only positional advantage. The disdain that almost all forms of doping have earned in sports derives largely from the combination of increased risk with the quest for positional advantage. Many sports involve significant risks, which we accept as a necessary part of the contest, but when an athlete uses steroids or EPO to gain an advantage over others, we object. One competitor's use of drugs puts pressure on everyone to do the same. In the end, no one is better off, and all are placed at greater risk. Many years ago the ecologist Garrett Hardin identified what he called the "tragedy of the commons."[47] The tragedy arises when grazing land is held in common. In these circumstances, a herder may find it rational to introduce more animals, as long as he thinks no one else will do so, since the slight reduction in fodder per animal is more than offset by his larger herd. But when all the herders, thinking the same way, follow suit, no one does any better overall, since there is only the same, limited amount of grass. The result is that all the cattle become undernourished, the commons becomes overgrazed, the grass withers, and tragedy ensues. Although sports doping raises other issues, the basic dynamics are the same. When anyone pursues personal advantage in the hope that others will not do so, a process sets in that ends by damaging everyone. Hardin not only diagnosed the problem, he offered a solution: "mutual coercion mutually agreed upon." Because the temptation in these circumstances to seek personal gain is so great, competitors usually cannot rely on good faith and trust. They must agree to set up an authority with the power to prohibit and punish cheating.

Some requests for gene enhancements, like sports doping, could produce a tragedy of the commons. As I hypothesized earlier, parents seeking a sports champion might try to have a child with an elevated red blood cell function. At its extreme, this request could significantly increase the child's risk of heart disease. Once many other parents started doing the same thing, the result would be no competitive advantage for anyone—bought at the price of increased health risks for all.

Educating parents would help avoid the worst excesses of gene enhancement. Medical professionals, journalists, and teachers should make clear how risky and futile the pursuit of enhancements aimed at achieving superiority could be. Everything should be done to appeal to parents' instinctive desire to protect their children from harm. But since parents are also tempted to move their children forward in life's competitive race, there is room for Hardin's "mutual coercion mutually agreed upon." Mutual coercion could take the form of professional standards of care that prohibit reprogenetic doctors from offering these enhancements and that provide a basis for wrongful-life suits against both overzealous parents and those who abet them. If the problem gets out of hand, prohibitory legislation, though very difficult to enforce, is also an option. India and China have both passed laws against sex selection with mixed results. It is always better to reduce the motivation for enhancement requests than to criminalize enhancement.

Not all requests for enhancements providing some benefits for a child have this problem of positional advantage. For example, if many parents strive to ensure that their children are more nearly normal in height or weight, excess and tragedy need not occur. Furthermore, even if better eyesight, reading ability, or IQ can aid in competitive sports or college admissions, they are also generally useful in life. These enhancements may even be socially beneficial. Who doesn't want a smarter doctor, a surgeon with improved hand-eye coordination, or a researcher with sharper analytical skills? The challenge in the future is to distinguish gene enhancements that confer only positional advantage and are harmful in the long term from those that have intrinsic value and that, on balance, are worth pursuing.

Fourth Guideline: Genetic interventions should not reinforce or increase unjust inequality and discrimination, economic inequality, or racism. I borrow this guideline from the bioethicist Julian Savulescu, but I offer it with some reservation.[48] I have already suggested the worrisome possibility that gene enhancements will widen the gap between the haves and the have-nots. Enhancements could further privilege the fortunate, and if they do so, we should think of effective ways of either controlling them or increasing access to them. Speaking of technologies that could enhance brain function, the bioethicist Arthur Caplan observes that "it is certainly possible—in fact, probable—that if nothing were done to ensure access to brain-enhancing technologies, inequities would arise." But he adds that the best solution "is to provide fair access . . . not to do away with the idea of improvement."[49] Properly handled, cognitive enhancements could also provide new and powerful ways of remedying some underlying problems, like persistent reading disabilities, that perpetuate economic failure.

Some enhancements could have the effect of reinforcing discrimination and racism. These include the use of prenatal gene modifications by African-American parents to lighten the color of their child's skin or straighten their child's hair; by East Asians to reduce the salience of the distinctive epicanthic fold that marks the eyes in order to produce a more "Western" appearance in their child; or by homophobic parents to avoid the birth of a gay child.

Each of these decisions has some tendency to reinforce discrimination. When black or Asian parents try to conform to white society's notions of beauty, they reinforce prejudice against dark skin and "unround" eyes. Those who exhibit such features, or who resist changing them, could find it harder to flourish in a society where these prejudices are increasingly widespread and where they are in a shrinking minority. Similarly, if parents use prenatal genetic means to reduce the chance that their child will be gay, they could contribute in various ways to the discrimination suffered by homosexual people. By avoiding the birth of a gay child they send the message that being gay is not desirable. This is true even if the parents are supportive of gay rights but are reluctant to visit the added burdens of a homosexual identity on their child. In addition, by reducing the size of the gay community, parents who make this choice diminish that community's political clout. In the early 1990s, when the NIH researcher Dean Hamer announced the discovery of a marker on the X chromosome possibly linked to a gene variant contributing to male homosexuality, people in the gay community voiced differing opinions.[50] Some celebrated the finding that homosexuality is not a choice but an inborn tendency. Others feared that the use of genetic technology would decimate the gay community. Headlines predicting "gene-ocide" appeared in some publications, and plays and films like *Twilight of the Golds* depicted how even progressive parents might be tempted to use this technology to avoid the birth of a gay child.[51] Early in 2007, the president of the Southern Baptist Theological Seminary raised the possibility of a genetic or other biological cause of homosexuality and urged conservative Christians to consider using biotechnology to eradicate the condition.[52]

The likelihood of reinforcing prejudice and unjust inequality is a serious concern. Nevertheless, I believe that it shouldn't lead us to panic and impose overly restrictive legal bans. For one thing, parents have a right to try to reduce the social burdens facing their offspring. Generations of Jewish parents have offered nose jobs to their adolescent children. This decision, for parent and child alike, is intensely personal, and we should be wary of imposing social restrictions on it. For another, there is an ebb and flow in social stereotypes. One black generation's quest for "relaxed" (long and straight) hair is replaced by another's preference for an Afro look. In the inevitable pendulum swings of culture, conformity to majoritarian preferences often gives way to minority pride. Before we introduce repressive and often useless legislation, it is worth waiting to see whether these shifts in taste are not largely self-corrective.

Our ability to use reprogenetics to shape our bodies will surely lead to bouts of conformism, but it may also bring about new expressions of diversity and uniqueness. Dean Hamer reminds those who criticize his research on the "gay gene" that for every straight couple trying to prevent the birth of a gay child there is a gay couple that could use this technology to have a child like themselves.[53] A related phenomenon has emerged as the amount of genetic information has increased: the deliberate effort by people with a disability, be it deafness or dwarfism, to use PGD to have children "like themselves."[54] Such choices poses difficult new ethical questions, in part because they appear to violate my first guideline, that genetic interventions should always be aimed at what is reasonably in the child's best interests. Nevertheless, they illustrate the point that reprogenetic technology may foster not just conformism and homogeneity but also new, and sometimes disturbing, expressions of freedom and diversity.

I began this book by stating that humankind is about to embark on the deliberate control of its own evolution. Although human germline gene modification remains on the distant horizon, advances in technology are inevitably moving it from the merely imaginable to the possible. Science fiction is becoming science reality. Whether in ten, twenty, or thirty years, the first deliberate inheritable modifications of human genes will take place. Enhancements will soon complement therapies. From that moment on, the pace of change will accelerate, and we will have fully entered the era of directed human evolution.

Let me pursue an example of change presented earlier. In 2005, a committee of Britain's House of Commons challenged the HFEA's refusal to permit sex selection for purposes of family balancing, indicating that the ban might soon be eased. Then, in late May 2006, the HFEA announced that it was going to permit in vitro fertilization patients to use preimplantation genetic diagnosis to avoid having children with mutations in the BRCA1 and BRCA2 genes. People who carry these mutations have a significantly higher risk of breast, ovarian, or prostate cancer than those without the mutations. Until recently, the HFEA prohibited such testing on the grounds that some people with these mutations never get cancer and that even those that do usually incur the disease only later in life. In the HFEA's thinking, using PGD for BRCA1 and BRCA2 testing was less like preventing disease and more like trying to have the perfect child. The HFEA's chairwoman at the time of the announcement, Suzi Leather, echoed the authority's previous opposition when she stated that the decision was "not about opening the door to wholesale genetic testing." Rather, genetic tests would be available to the minority of people with a clear history of cancer in the family.[55]

The critics spoke up. Francis Collins, director of the U.S. National Human Genome Research Institute in Bethesda, Maryland, whose organization helped fund the work of the Human Genome Project that sped identification of the BRCA genes, said, "I'm not entirely comfortable because of the concerns about the whole spectrum, from very severe diseases to what are essentially traits. There is no bright line along that spectrum." Disability campaigners added their voices, saying that the HFEA decision smacked of eugenics. Josephine Quintavalle of the U.K. lobby group Comment on Reproductive Ethics complained, "We are concerned that people are eliminating embryos, whether they have cancer or not." She said that research efforts should be concentrated on cancer cures, not destroying affected embryos.[56]

Despite these concerns and objections, the HFEA's reversal is a sign of things to come. It recognizes that parents have a right to use the new reprogenetic technologies to ease their worries about a child, to free the child from needless burdens, and to give the child the best start in life. The worriers and critics are correct to believe that the decision is another milestone on the path to more extensive interventions, including, somewhere down the line, gene enhancements. But they are wrong to regard these developments only with fear. Having vastly expanded our control over the world around us, our species is now rapidly developing the ability to alter the world within. The question is not *whether* we will do this but *when* and *how*.

In discussing sports enhancement and gene doping, I indicated people's tendency to alight on a single consideration in assessing a practice. Some will criticize such boosts as "unnatural." Others raise questions about fairness or justice. Still others stress risks or worry about the way the intervention may detract from the interest or spirit of the sport. Each of these considerations is a valid concern, but I pointed out that no single one should be the sole basis of decision. Most sports raise safety concerns, so safety is only one factor in our thinking. Decisions about fairness or the integrity of a sport incorporate other values and must be balanced against such mundane considerations as our ability to enforce any rules we make.

The same approach applies to gene modification and enhancement generally. Here, too, we need a multifaceted decision process. We should not reject an enhancement just because it is unnatural or potentially unfair or unjust. Instead, we have to assess and balance safety, fairness, justice, the enforceability of any prohibitions, and all the various impacts on parents, children, and society. Individuals and society are beginning this thinking now. In this book, I have tried to further it, but I have made only a start. These questions will be leading items on the bioethics agenda of the twenty-first century.

To grow morally and spiritually has always been a human aim. This is the goal of Judaism, Christianity, Islam, and nearly every other religion. The ethicist-theologian Ronald Cole-Turner reminds us that sometimes *how* we achieve a goal is as important as the goal itself. We cannot simply replace mental and spiritual growth with drugs or biotechnologies that promise to make us smarter or wiser.[57] But neither do human beings live by spirit alone; we are psychosomatic beings, psychological and physical wholes. Why should we rule out changes to our bodies and our genes as we pursue the goal of human advancement? Scientists have begun to suggest that evolutionary increases in our cognitive capacities may have contributed to the striking advances in civilization seen in relatively recent human history. Might directed changes at the genetic level play a role in fostering our continuing cultural, moral, and spiritual growth?

In *Consilience,* the biologist Edward O. Wilson observes that "genetic evolution is about to become conscious and volitional, and usher in a new epoch in the history of life." Wilson believes—and hopes—that we will be "conservative" in our choices, always striving to preserve what makes us human.[58] I agree with Wilson's caution. But I also believe that our biology is far from perfect, that there are new mental and physical potentialities we can develop, and that what makes us human is open to continuous discovery. By moving responsibly into this emerging era of genetic choice, we can incorporate gene technology into the ongoing human adventure.

Notes

Introduction

1. Lee M. Silver, *Remaking Eden* (New York: Avon Books, 1997).
2. National Human Genome Research Institute, "International Consortium Completes Human Genome: All Goals Achieved; New Vision for Genome Research Unveiled," Bethesda, MD, April 14, 2003, available online at http://www.genome.gov/11006929.
3. Robert F. Service, "The Race for the $1,000 Genome," *Science* 311 (March 17, 2006), 1544–1546; Nicholas Wade, "New Methods to Sequence DNA Promise Vastly Lower Costs," *New York Times,* August 9, 2005; Wade, "The Quest for the $1,000 Human Genome," *New York Times,* July 18, 2006; Wade, "$10 Million Prize Set Up for Speedy DNA Decoding," *New York Times,* October 5, 2006; and John A. Robertson, "The $1,000 Genome: Ethical and Legal Issues in Whole Genome Sequencing of Individuals," *American Journal of Bioethics* 3/3 (Summer 2003), W-IF1. The declining cost of genomic sequencing has already made it possible to sequence the complete genome of famous individuals, raising novel privacy questions. See Eliot Marshall, "Sequencers of a Famous Genome Confront Privacy Issues," *Science* 315 (March 30, 2007), 1780.
4. President's Council on Bioethics, *Beyond Therapy: Biotechnology and the Pursuit of Happiness* (Washington, DC: President's Council on Bioethics, 2003), available online at http://www.bioethics.gov/reports/beyondtherapy/index.html.
5. A record of critical remarks about biomedicine was gathered by Ruth Macklin, "The New Conservatives in Bioethics: Who Are They and What Do They Seek?" *Hastings Center Report* 36/1 (2006), 34–43. For additional assessments of the council's work, see Eric M. Meslin, "The President's Council: Fair and Balanced?" *Hastings*

Center Report 34/2 (2004), 6–8; and Ronald M. Green "For Richer or Poorer? Evaluating the President's Council on Bioethics," *HEC Forum,* 18/2 (2006), 108–124.

6. Daniel Callahan, *Setting Limits: Medical Goals in an Aging Society* (New York: Simon and Schuster, 1987); Callahan, *What Kind of Life: The Limits of Medical Progress* (New York: Simon and Schuster, 1990); Callahan, *The Troubled Dream of Life* (Washington, DC: Georgetown University Press, 2000). See also Callahan, "End of Life Care," *Bi-Annual Devos Medical Ethics Colloquy,* March 2, 2006.

7. Bill McKibben, *Enough: Staying Human in an Engineered Age* (New York: Henry Holt, 2003).

8. UNESCO, Universal Declaration on the Human Genome and Human Rights, November 11, 1997, Article 1, available online at http://portal.unesco.org/shs/en/ev.php-URL_ID=1881&URL_DO=DO_TOPIC&URL_SECTION=201.html.

9. Ursula K. Le Guin offers a similar apocalyptic vision in the short story "Porridge on Islac," in Le Guin, *Changing Planes* (New York: Ace Books, 2003).

10. William Samuelson and Richard Zeckhauser, "Status Quo Bias in Decision Making," *Journal of Risk and Uncertainty* 1 (1988), 10.

11. Ralph Brave, "James Watson Wants to Build a Better Human," AlterNet, May 29, 2003, available online at http://www.alternet.org/story/16026/; Gregory Stock, *Redesigning Humans: Our Inevitable Future* (Boston: Houghton Mifflin, 2002).

12. Gina Kolata, "The New Age: So Big and Healthy Grandpa Wouldn't Even Know You," *New York Times,* July 30, 2006; Robert W. Fogel, "Changes in the Physiology of Aging during the Twentieth Century," NBER Working Paper No. 11233, March 2005, available online at http://www.nber.org/papers/w11233.

13. For an excellent overview of these developments, see Nicholas Wade, *Before the Dawn: Recovering the Lost History of Our Ancestors* (New York: Penguin, 2006).

14. Alison P. Galvani and Montgomery Slatkin, "Evaluating Plague and Smallpox as Historical Selective Pressures for the CCR5-Δ32 HIV-Resistance Allele," *Proceedings of the National Academy of Sciences of the United States of America* 100/25 (2003), 15276–15279.

15. Wade, *Before the Dawn,* 134–138; Sarah A. Tishkoff et al., "Convergent Adaptation of Human Lactase Persistence in Africa and Europe," *Nature Genetics,* published online on December 10, 2006; Nicholas Wade, "Lactose Tolerance in East Africa Points to Recent Evolution," *New York Times,* December 11, 2006.

16. Patrick D. Evans et al., "Microcephalin, a Gene Regulating Brain Size, Continues to Evolve Adaptively in Humans," *Science* 309 (September 9, 2005), 1717–1720; Nitzan Mekel-Bobrov et al., "Ongoing Adaptive Evolution of ASPM, a Brain Size Determinant in *Homo sapiens,*" *Science* 309 (September 9, 2005), 1720–1722.

17. Michael Balter, "Are Human Brains Still Evolving? Brain Genes Show Signs of Selection," *Science* 309 (September 9, 2005), 1662–1663.

18. Benjamin F. Voight et al., "A Map of Recent Positive Selection in the Human Ge-
    nome," *PLoS Biology* 4/3 (March 7, 2006), available online at http://biology
    .plosjournals.org/perlserv/?request=get-document&doi=10.1371/journal.pbio
    .0040072 and http://web.ebscohost.com/ehost/detail?vid=3&hid=107&sid=
    b915b814-46f6-48ef-bca3-460544c56060%40sessionmgr106.

19. Antonio Regalado, "Scientist's Study of Brain Genes Sparks a Backlash—Dr. Lahn
    Connects Evolution in Some Groups to IQ; Debate on Race and DNA," *Wall Street
    Journal,* June 16, 2006.

20. The role of natural selection versus purely demographic factors in the prevalence
    of these genes has recently been debated in two articles: Mathias Currat et al.,
    "Comment on 'Ongoing Adaptive Evolution of ASPM, a Brain Size Determinant in
    Homo sapiens' and 'Microcephalin, a Gene Regulating Brain Size, Continues to
    Evolve Adaptively in Humans,'" *Science* 313 (July 14, 2006), 172a; and Nitzan
    Mekel-Bobrov et al., "Response to Comment on 'Ongoing Adaptive Evolution of
    ASPM, a Brain Size Determinant in Homo sapiens' and 'Microcephalin, a Gene
    Regulating Brain Size, Continues to Evolve Adaptively in Humans,'" *Science* 313
    (July 14, 2006), 172b. More recently, Bruce Lahn and his colleagues have argued
    that the presence of the variant Microcephalin gene may be due to a rare inter-
    breeding event (or events) between the modern human and the Neanderthal lin-
    eages. See Patrick D. Evans, Nitzan Mekel-Bobrov, Eric J. Vallender, Richard R. Hud-
    son, and Bruce T. Lahn, "Evidence That the Adaptive Allele of the Brain Size Gene
    Microcephalin Introgressed into *Homo* from an Archaic *Homo* Lineage," *Proceedings
    of the National Academy of Sciences of the United States of America* 103/48 (2006),
    18178–18183.

21. Kean Birch, "Beneficence, Determinism and Justice: An Engagement with the Ar-
    gument for the Genetic Selection of Intelligence," *Bioethics* 19/1 (2005), 12–28.

22. Erik Parens, "Genetic Differences and Human Identities: On Why Talking about Be-
    havioral Genetics Is Important and Difficult," *Hastings Center Report Special Supple-
    ment* 34/1 (2004), S29.

23. Lee M. Silver, *Challenging Nature* (New York: HarperCollins, 2006), 347–348.

1. Creating the Superathlete

1. E. R. Barton-Davis, D. I. Shoturma, A. Musarò, N. Rosenthal, and H. L. Sweeney, "Viral Mediated Expression of Insulin-Like Growth Factor I Blocks the Aging-Related
   Loss of Skeletal Muscle Function," *Proceedings of the National Academy of Sciences of
   the United States of America* 95/26 (1998), 15603–15607; H. Lee Sweeney, "Gene
   Doping," *Scientific American,* July 2004, 63–69.

2. Antonio Musarò, Karl McCullagh, Angelika Paul, Leslie Houghton, Gabriella Do-

browolny, Mario Molinaro, Elisabeth R. Barton, H. Lee Sweeney, and Nadia Rosenthal, "Localized Igf-1 Transgene Expression Sustains Hypertrophy and Regeneration in Senescent Skeletal Muscle," *Nature Genetics* 27 (2001), 195–200. Recently a team of researchers at Dartmouth College identified a gene that regulates glycogen storage, a factor in both muscle-wasting diseases and athletic performance. The team produced transgenic, genetically engineered mice with an activated form of the gene. These mice were able to retain high levels of muscle function without exercise. See Laura Barre et al., "A Genetic Model for the Chronic Activation of Skeletal Muscle AMP-Activated Protein Kinase Leads to Glycogen Accumulation," *American Journal of Physiology, Endocrinology and Metabolism,* published online November 2006 at http://ajpendo.physiology.org/cgi/content/abstract/00369.2006v1?maxtoshow=&HITS=10&hits=10&RESULTFORMAT=1&author1=witters%2C+L&andorexacttitle=and&andorexacttitleabs=and&andorexactfulltext=and&searchid=1&FIRSTINDEX=0&sortspec=relevance&resourcetype=HWCIT.

3. Philip R. Reilly, *The Strongest Boy in the World: How Genetic Information Is Reshaping Our Lives* (Cold Spring Harbor, NY: Cold Spring Harbor Laboratory Press, 2006), 6.

4. N. Yang et al., "ACTN3 Genotype Is Associated with Human Elite Athletic Performance," *American Journal of Human Genetics* 73, 627–631.

5. Theodore Friedmann, conversation with the author, January 9, 2006.

6. Craig Lord, "Drug Claim Could Be a Bitter Pill," *Times Online* (timesonline.co.uk), March 2, 2005, available online at http://www.ergogenics.org/ddrroids2.html.

7. Jesper L. Andersen, Peter Schjerling, and Bengt Saltin, "Muscle, Genes and Athletic Performance," *Scientific American,* September 2000, 48–55.

8. Kevin Joseph, *The Champion Maker* (Bloomington, IN: Unlimited Publishing, 2005), 244.

9. Ibid., 246.

10. I draw the term "über-athletes" from John Entine, "The Coming of the Über-Athlete," Salon.com, March 21, 2002, available online at http://www.salon.com/news/sports/2002/03/21/genes/print.html.

11. Gina Kolata, "Race to the Swift? Not Necessarily," *New York Times,* July 18, 2006.

12. Gina Kolata, "Live at Altitude? Sure. Sleep There? Hmm," *New York Times,* July 26, 2006.

13. Andy Miah, *Genetically Modified Athletes: Biomedical Ethics, Gene Doping and Sport* (London: Routledge, 2004), 25.

14. Christie Aschwanden, "Gene Cheats," *New Scientist,* January 15, 2000, 24, available online at http://www.archway.ac.uk/Activities/Departments/SHHP/downloads/epo/Genecheats/genecheats.html.

15. Michael Specter, "The Long Ride: How Did Lance Armstrong Manage the Greatest Comeback in Sports History?" *New Yorker,* July 15, 2002, available online at http://www.michaelspecter.com/ny/2002/2002_07_15_lance.html.

16. Quoted in Stefan Lovgren, "Olympic Gold Begins with Good Genes, Experts Say," *National Geographic News,* August 20, 2004, available online at http://news.nationalgeographic.com/news/2004/08/0820_040820_olympics_athletes.html.

17. See her biography, Linda Armstrong Kelly with Joni Rodgers, *No Mountain High Enough: Raising Lance, Raising Me* (New York: Broadway Books, 2005).

18. Quoted in Lovgren, "Olympic Gold Begins with Good Genes."

19. J. Savulescu, B. Foddy, and M. Clayton, "Why We Should Allow Performance Enhancing Drugs in Sport," *British Journal of Sports Medicine* 36/6 (December 2004), 667.

20. Rolf Jarmann, quoted in Elmar Wagner, "Aus der Perspektive des Sportlers—eine Innensicht," in Michael Gamper, Jan Mühlethaler, and Felix Reidhaar, eds., *Doping. Spitzensport als gesellschaftliches Problem* (Zurich: NZZ Verlag, 2000), 35.

21. Savulescu, Foddy, and Clayton, "Why We Should Allow Performance Enhancing Drugs in Sport," 666.

22. CBS Evening News, May 28, 2006.

2. How Will We Do It?

1. Susan Sample, "Scientist Profile: Mario Capecchi, Ph.D.," Genetic Science Learning Center, University of Utah, 2007, available online at http://gslc.genetics.utah.edu/features/capecchi/.

2. For an account of David Vetter's ordeal, see Leroy Walters and Julie Gage Palmer, *The Ethics of Human Gene Therapy* (New York: Oxford University Press, 1997), xiii–xvi.

3. S. Hacein-Bey-Abina et al., "LMO2-Associated Clonal T Cell Proliferation in Two Patients after Gene Therapy for SCID-X1," *Science* 302 (October 17, 2003), 415–419; Jocelyn Kaiser, "Panel Urges Limits on X-SCID Trials," *Science* 307 (March 11, 2005), 1544.

4. Christopher Baum et al., "Chance or Necessity? Insertional Mutagenesis in Gene Therapy and Its Consequences," *Molecular Therapy* 9/1 (January 2004), 5–13.

5. Sheryl Gay Stolberg, "The Biotech Death of Jesse Gelsinger," *New York Times Magazine,* November 28, 1999.

6. One article is Theodore Friedmann's "Overcoming the Obstacles," *Scientific American,* June 1997, 96–101.

7. Zinc Finger Consortium, "Scientific Background," available online at http://www .zincfingers.org/scientific-background.htm; Jocelyn Kaiser, "Putting the Fingers on Gene Repair," *Science* 310 (December 23, 2005), 1894–1896. For another approach, see Kenneth Culver, "Correction of Chromosomal Point Mutations in Human Cells with Bifunctional Oligonucleotides," *Nature Biotechnology* 17 (October 1999), 989–963.

8. Deborah O. Co et al., "Generation of Transgenic Mice and Germline Transmission of a Mammalian Artificial Chromosome Introduced into Embryos by Pronuclear Microinjection," *Chromosome Research* 8 (2000), 183–191; S. Vanderbyl et al., "Transfer and Stable Transgene Expression of a Mammalian Artificial Chromosome into Bone-Marrow-Derived Human Mesenchymal Stem Cells," *Stem Cells* 22/3 (2004), 324–333, available online at http://stemcells.alphamedpress.org/cgi/ reprint/22/3/324.pdf.

9. Bhavani G. Pathak, "Scientific Methodologies to Facilitate Inheritable Genetic Modifications in Humans," in Audrey R. Chapman and Mark S. Frankel, eds., *Designing Our Descendants: The Promises and Perils of Genetic Modifications* (Baltimore, MD: Johns Hopkins University Press, 2003), 55–67; Theodore Friedmann, "Approaches to Gene Transfer to the Mammalian Germ Line," in Chapman and Frankel, *Designing Our Descendants,* 44.

10. Sabin Russell, "Of Mice and Men: Striking Similarities at the DNA Level Could Aid Research," *San Francisco Chronicle,* December 5, 2002, available online at http://www.sfgate.com/cgi-bin/article.cgi?f=/c/a/2002/12/05/ MN153329.DTL&type=science. For a different description of the similarities and differences in these genomes, see Alec MacAndrew, "What Does the Mouse Genome Draft Tell Us about Evolution?" Mouse Genome Homepage, March 1, 2003, available online at http://www.evolutionpages.com/Mouse%20genome %20home.htm; Chimpanzee Sequencing and Analysis Consortium, "Initial Sequence of the Chimpanzee Genome and Comparison with the Human Genome," *Nature* 436 (2005), 69–87; NIH News, "Genome Comparison Finds Chimps, Humans Very Similar at the DNA Level," August 31, 2005, available online at http://www.genome.gov/15515096. For a discussion of the complexity of evaluating the meaning of quantitative differences in genomic similarities across closely related species, see Jonathan Marks, *What It Means to Be 98% Chimpanzee: Apes, People, and Their Genes* (Berkeley: University of California Press, 2002), chaps. 1–2.

11. H. Watanabe et al., "DNA Sequence and Comparative Analysis of Chimpanzee Chromosome 22," *Nature* 429 (2004), 382–388; Jean Weissenbach, "Genome Sequencing: Differences with the Relatives," *Nature* 429 (2004), 353–355.

12. F. Vargha-Khadem et al., "Praxic and Nonverbal Cognitive Deficits in a Large Family with a Genetically Transmitted Speech and Language Disorder," *Proceedings of the National Academy of Sciences of the United States of America* 92/3 (1995), 930-933; K. E. Watkins, N. F. Dronkers, and F. Vargha-Khadem, "Behavioural Analysis of an Inherited Speech and Language Disorder: Comparison with Acquired Aphasia," *Brain* 125 (2002), 452-464; Wolfgang Enard et al., "Molecular Evolution of FOXP2, a Gene Involved in Speech and Language," *Nature* 418 (2002), 869-872. For a discussion of these cumulative findings, see Henry Gee, *Jacob's Ladder: The History of the Human Genome* (New York: W. W. Norton, 2004), 222-225.

13. Elizabeth Pennisi, "Genomes Throw Kinks in Timing of Chimp-Human Split," *Science* 312 (May 19, 2006), 986-987.

14. Some biologists argue that precursors to morality are evident in primate populations, although they have been significantly developed as a result of the reasoning abilities of our species. See Nicholas Wade, "Scientist Finds the Beginnings of Morality in Primate Behavior," *New York Times*, March 20, 2007.

15. See GeneTests, available online at http://www.geneclinics.org/.

16. Treating the cell donor embryo would probably require temporarily freezing that embryo after utilization of single cell blastomere biopsy as a new technology for stem cell derivation. For a description of this technology, see Irina Klimanskaya et al., "Human Embryonic Stem Cell Lines Derived from Single Blastomeres," *Nature* 444 (November 23, 2006), 481-485.

17. André C. Van Steirteghem, "Outcomes of Assisted Reproductive Technology," *New England Journal of Medicine* 338/3 (January 15, 1998), 194-195; Claire Ainsworth, "The Stranger Within," *New Scientist* 180/2421 (November 15, 2003), 34.

18. Patrick W. Dunne and Jorge A. Piedrahita, "Genetic Modification and Cloning in Mammals," in Jose Cibelli, Robert P. Lanza, Keith H. S. Campbell, and Michael D. West, eds., *Principles of Cloning* (New York: Academic Press, 2002), 227-246.

19. "Backing for Baby Cloning to Beat Disease," *Daily Telegraph,* June 5, 2006, available online at http://www.telegraph.co.uk/news/main.jhtml?xml=/news/2006/06/05/nclone05.xml&sSheet=/news/2006/06/05/ixuknews.html.

20. M. Sato, A. Ishikawa, and M. Kimura, "Direct Injection of Foreign DNA into Mouse Testis as a Possible In Vivo Gene Transfer System via Epididymal Spermatozoa," *Molecular Reproduction and Development* 61 (2002), 49-56.

21. Sarah Wildman, "Stop Time," *New York Magazine,* October 17, 2005; Sally Wadyka, "For Women Worried about Fertility, Egg Bank Is a New Option," *New York Times,* September 21, 2004; Caroline Ryan, "Egg Freezing Boosts Baby Chances," BBC News, Prague, June 19, 2006, available online at http://news.bbc.co.uk/2/hi/health/5095096.stm.

3. Drawing Lines

1. This schema was first suggested in Leroy Walters, "Ethical Issues in Human Gene Therapy," *Journal of Clinical Ethics* 2/4 (1991), 267–274.

2. For a discussion of the special risks of any type of gene transfer research, see Jonathan Kimmelman, "Recent Developments in Gene Transfer: Risk and Ethics," *British Medical Journal* 330 (January 8, 2005), 79–82.

3. Esmail D. Zanjani and W. French Anderson, "Prospects for in Utero Human Gene Therapy," *Science* 285 (September 24, 1999), 2084–2088.

4. Michael Boylan and Kevin E. Brown, *Genetic Engineering: Science and Ethics on the New Frontier* (Upper Saddle River, NJ: Prentice Hall, 2001), 157–159.

5. For a discussion of whether medical care should ever embrace pure enhancements and whether a therapy-enhancement distinction can be maintained in the area of gene modifications, see David B. Resnik, "In Pursuit of Perfect People: The Ethics of Enhancement; The Moral Significance of the Therapy-Enhancement Distinction in Human Genetics," *Cambridge Quarterly of Healthcare Ethics* 9 (2000), 365–367.

6. Discussions of the controversy include John Lantos, Mark Siegler et al., "Ethical Issues in Growth Hormone Therapy," *JAMA: The Journal of the American Medical Association* 261 (1989), 1020–1024; D. B. Allen and N. C. Fost, "Growth Hormone Therapy for Short Stature: Panacea or Pandora's Box?" *Journal of Pediatrics* 117 (1990), 16–21; and Allen and Fost, "hGH for Short Stature: Ethical Issues Raised by Expanded Access," *Journal of Pediatrics* 144 (May 2004), 648–652; N. Dov Fox, "Human Growth Hormone and the Measure of Man," *New Atlantis*, no. 7 (Fall 2004–Winter 2005), 75–87. For a discussion of the complexities of pursuing research on hGH, see Carol A. Tauer, "The NIH Trials of Growth Hormone for Short Stature," *IRB: Ethics and Human Research* 16 (1994), 1–9; Thomas H. Murray, *The Worth of a Child* (University of California Press: Berkeley, 1996), 87–93.

7. Urs Eiholzer, Fritz Haverkamp, and Linda Voss, eds., *Growth, Stature and Psychosocial Well-Being* (Seattle: Hogrefe and Huber, 1999). See also Stephen S. Hall, "The Short of It," *New York Times Magazine,* October 16, 2005.

8. Bernard Gert, Charles M. Culver, and K. Danner Clouser point to the role played in our conception of illness or disease by two independent ideas: the abnormality of a bodily condition and its tendency to lead to an increased risk of suffering. In contrast to the authors of a host of previous accounts, they do not stress our need to understand the condition's biological causation or underlying malfunction. See Gert, Culver, and Clouser, *Bioethics: A Return to Fundamentals* (New York: Oxford University Press, 1997), chap. 5.

9. In November 2006 the FDA lifted a fourteen-year ban on the use of silicone implants in women other than those who needed to replace a failed implant or who desired

reconstruction after cancer or injuries. For brief reviews of the lengthy controversy over silicone breast implants, see Steven Reinberg, "FDA OKs Return of Silicone Breast Implants," *MSN Health and Fitness,* November 17, 2006, available online at http://health.msn.com/womenshealth/ArticlePage.aspx?cp-documentid= 100148882; and MSNBC, "Ban on Silicone Breast Implants Lifted," November 17, 2006, available online at http://www.msnbc.msn.com/id/15770935/.

10. Eric T. Juengst, "Can Enhancement Be Distinguished from Prevention in Genetic Medicine?" *Journal of Medicine and Philosophy* 22 (1997), 125–142. Juengst has observed that prevention of disease by genetic interventions can be pursued in several further ways; see Juengst, "Prevention and the Goals of Genetic Medicine," *Human Gene Therapy* 8/12 (December 1996), 1595–1605. See also Juan Manuel Torres, "On the Limits of Enhancement in Human Gene Transfer: Drawing the Line," *Journal of Medicine and Philosophy* 22 (1997), 43–53.

11. Judy C. Chang, Lin Ye, and Yuet Wai Kan, "Correction of the Sickle Cell Mutation in Embryonic Stem Cells," *Proceedings of the National Academy of Sciences of the United States of America* 103/4 (2006), 1036–1040. See also Allyson Cole-Strauss et al., "Correction of the Mutation Responsible for Sickle Cell Anemia by an RNA-DNA Oligonucleotide," *Science* 273 (September 6, 1996), 1386–1389.

12. Edward M. Berger, "Ethics of Gene Therapy," in Bernard Gert et al., *Morality and the New Genetics: A Guide for Students and Health Care Providers* (Boston: Jones and Bartlett, 1996), 209–223. This concern assumes that germline gene modifications cannot be made reversible or subject to the recipient generations' choice of whether they should be passed on.

13. Nelson A. Wivel and Leroy Walters, "Germ-Line Gene Modification and Disease Prevention: Some Medical and Ethical Perspectives," *Science* 262 (October 22, 1993), 533–538.

14. Pope John Paul II, "Biological Research and Human Dignity," *Origins* 12/21 (November 4, 1982), 343; Pope John Paul II, "The Ethics of Genetic Manipulation: Address to the World Medical Association," *Origins* 13/23 (November 17, 1983), 388.

15. Rebecca Dresser argues that couples seeking to avoid prenatal selection "would still face choices about whether to initiate or continue a pregnancy when tests suggested that the resulting child would be unhealthy," but she underestimates some parents' willingness to forge ahead in order to avoid embryo selection or abortion. See Dresser, "Designing Babies: Human Research Issues," *IRB: Ethics and Human Research* 26/5 (2004), 4.

16. Nicholas Balakar, "Ugly Children May Get Parental Short Shrift," *New York Times,* May 3, 2005.

17. Elaine Hatfield and Susan Sprecher, *Mirror, Mirror: The Importance of Looks in Everyday Life* (Albany, NY: State University of New York Press, 1986).

18. Ibid., 97-98. P. L. Benson, S. A. Karabenick, and R. M. Lerner, "Pretty Pleases: The Effects of Physical Attractiveness, Race, and Sex, on Receiving Help," *Journal of Experimental Social Psychology* 12 (1976), 409-415.

19. Hatfield and Sprecher, *Mirror, Mirror*, 90; H. Sigall and N. Ostrove, "Beautiful but Dangerous: Effects of Offender Attractiveness and the Nature of the Crime on Juridic Judgment," *Journal of Personality and Social Psychology* 31 (1975), 410-414; M. R. Solomon and J. Schopler, "The Relationship of Physical Attractiveness and Punitiveness: Is the Linearity Assumption Out of Line," *Personality and Social Psychology Bulletin* 4 (1978), 483-486.

20. Hatfield and Sprecher, *Mirror, Mirror*, 36; Sigall and Ostrove, "Beautiful but Dangerous."

21. Hatfield and Sprecher, *Mirror, Mirror*, chap. 3.

22. Ibid., 48-50.

23. R. Thornhill and S. Gangestad, "Human Facial Beauty: Averageness, Symmetry, and Parasite Resistance," *Human Nature: An Interdisciplinary Biosocial Perspective* 4 (1993), 237-269, quoting a study by J. Langlois, J. H. Roggman, and L. A. Roggman, "Attractive Faces Are Only Average," *Psychological Science* 1 (1990), 115-121.

24. This is reported in the investigative documentary *Frozen Angels*, an examination of the practices of egg donor programs in California. It was produced and directed by Eric Black and Frauke Sandig (Berlin: Umbrella Films, 2005).

25. Michael Balter, "Zebrafish Researchers Hook Gene for Human Skin Color," *Science* 310 (December 16, 2005), 1765-1766; Rebecca L. Lamason et al., "SLC24A5, a Putative Cation Exchanger, Affects Pigmentation in Zebrafish and Humans," *Science* 310 (December 16, 2005), 1782-1786.

26. Maria Capecchi, personal communication to the author, August 15, 2005.

27. Murray, *Worth of a Child*, 89.

28. Rick Weiss, "Gene Enhancements' Thorny Ethical Traits; Rapid-Fire Discoveries Force Examination of Consequences," *Washington Post*, October 12, 1997.

29. Allen and Fost, "Growth Hormone Therapy for Short Stature." For a higher, more recent cost estimate, see Fox, "Human Growth Hormone."

30. Andrew Postlewaite and Dan Silverman, "The Effect of Adolescent Experience on Labor Market Outcomes: The Case of Height," *Journal of Political Economy* 112/5 (October 2004), 1019-1053; see also Sheila M. Rothman and David J. Rothman, *The Pursuit of Perfection: The Promise and Perils of Medical Enhancement* (New York: Pantheon Books, 2001), 184.

31. Kristie M. Engemann and Michael T. Owyang, "So Much for That Merit Raise: The Link between Wages and Appearance," *Regional Economist*, April 2005, available

online at http://stlouisfed.org/publications/re/2005/b/pages/appearances
.html. The authors draw their data from Malcolm Gladwell, *Blink: The Power of Thinking without Thinking* (2005).

32. Hall, "The Short of It."

33. John S. Gillis, *Too Tall, Too Small* (Champaign, IL: Institute for Personality and Ability Testing, 1982), 21.

34. National Institute of Diabetes and Digestive and Kidney Diseases (NIDDK), "Understanding Adult Obesity," NIH Publication No. 01-3680, October 2001, available online at http://www.wrongdiagnosis.com/artic/understanding_adult_obesity _niddk.htm; World Health Organization, "Global Strategy on Diet, Physical Activity and Health: Obesity and Overweight," available online at http://www.who.int/ dietphysicalactivity/publications/facts/obesity/en/.

35 Hatfield and Sprecher, *Mirror, Mirror,* 214-215.

36. Christopher G. Bell, Andrew J. Walley, and Philippe Froguel, "The Genetics of Human Obesity," *Nature Reviews Genetics* 6 (March 2005), 221-234. For a slightly lower estimate of the heritability of obesity based on studies of twins, see Helen N. Lyon and Joel N. Hirschhor, "Genetics of Common Forms of Obesity: A Brief Overview," *American Journal of Clinical Nutrition* 82, 1 supplement (2005), 215S–217S.

37. Gina Kolata, "How the Body Knows When to Gain or Lose," *New York Times,* October 17, 2000.

38. Y. Zhang et al., "Positional Cloning of the Mouse Obese Gene and Its Human Homologue," *Nature* 372 (1994), 425-432; J. M. Friedman and J. L. Halaas, "Leptin and the Regulation of Body Weight in Mammals," *Nature* 395 (1998), 763-770; M. A. Pelleymounter et al., "Effects of the *obese* Gene Product on Body Weight Regulation in *ob/ob* Mice," *Science* 269 (July 28, 1995), 540-543; Alpha Diagnostic International, "Detail information of Tubby, Tub, TULP1, TULP2, Agouti, and AGRP and Mahogany Antibodies," available online at http://www.4adi.com/productflr/ tubbyflr.html.

39. Alan Herbert et al., "A Common Genetic Variant Is Associated with Adult and Childhood Obesity," *Science* 312 (April 14, 2006), 279-283.

40. Bell, Walley, and Froguel, "Genetics of Human Obesity," 231.

41. Kolata, "How the Body Knows When to Gain or Lose."

42. Elisabeth Rosenthal, "Europeans Find Extra Options for Staying Slim," *New York Times,* January 3, 2006; Robert Pear, "Obesity Surgery Often Leads to Complications, Study Says," *New York Times,* July 24, 2006.

43. John Travis, "Mouse Study Suggests Cancer Drugs Could Help Prematurely Aging Kids," *Science* 311 (February 17, 2006), 934-935.

44. Thomas Perls and Dellara Terry, "Understanding the Determinants of Exceptional Longevity," *Annals of Internal Medicine* 139 (2003), 445–449.

45. R. G. Cutler, "Evolution of Human Longevity and the Genetic Complexity Governing Aging Rate," *Proceedings of the National Academy of Sciences of the United States of America* 72/11 (1975), 4664–4668.

46. Dellara F. Terry et al., "Cardiovascular Advantages among the Offspring of Centenarians," *Journal of Gerontology: Medical Sciences* 58A/5 (2003), 425–431; Dellara F. Terry et al., "Cardiovascular Disease Delay in Centenarian Offspring," *Journal of Gerontology* 59A/4 (2004), 385–389.

47. T. T. Perls, L. M. Kunkel, and A. A. Puca, "The Genetics of Exceptional Human Longevity," *Journal of the American Geriatric Society* 50 (2002), 359–368. See also Andrzej Bartke et al., "Extending the Lifespan of Long-Lived Mice," *Nature* 414 (November 22, 2001), 412.

48. For a review of caloric restriction work in animals, see R. Weindruch et al., *The Retardation of Aging and Disease by Dietary Restriction* (Springfield, IL: Charles Thomas, 1998). See also Michael Mason, "One for the Ages: A Prescription That May Extend Life," *New York Times,* October 31, 2006.

49. Blanka Rogina et al., "Extended Life-Span Conferred by Cotransporter Gene Mutations in *Drosophila,*" *Science* 290 (December 15, 2000), 2137–2140; Blanka Rogina, Stephen L. Helfand, and Stewart Frankel, "Longevity Regulation by *Drosophila* Rpd3 Deacetylase and Caloric Restriction," *Science* 298 (November 29, 2002), 1745.

50. Michael Rose estimates that as many as five hundred genes may control aging in human beings. See Rose, *The Long Tomorrow: How Advances in Evolutionary Biology Can Help Us Postpone Aging* (New York: Oxford University Press, 2005), 500.

51. Michael Rose does not believe that enhanced longevity is best accomplished by genetic interventions. See his *Long Tomorrow,* 132.

52. Ya-Ping Tang, Eiji Shimizu, Gilles R. Dube, Claire Rampon, Geoffrey A. Kerchner, Min Zhuo, Guosong Liu, and Joe Z. Tsien, "Genetic Enhancement of Learning and Memory in Mice," *Nature* 401 (September 2, 1999), 63–69.

53. Tsien believes that the mice are not more sensitive to pain but are just able to learn to avoid it more quickly than other mice. See Y. P. Tang, E. Shimizu, and J. Z. Tsien, "Do 'Smart' Mice Feel More Pain or Are They Just Better Learners?" *Nature Neuroscience* 4/5 (2001), 453.

54. Howard Gardner, *Multiple Intelligences: The Theory in Practice* (New York: Basic Books, 1993).

55. Per Svenningsson et al., "Alterations in 5-HT1B Receptor Function by p11 in Depression-Like States," *Science* 311 (January 6, 2006), 77–80.

56. K. Lesch et al., "Association of Anxiety-Related Traits with a Polymorphism in the

Serotonin Transporter Gene Regulatory Region," *Science* 274 (November 29, 1996), 1527–1531; H. Kunugi et al., "Serotonin Transporter Gene Polymorphisms: Ethnic Differences and Possible Association with Bipolar Affective Disorder," *Molecular Psychiatry* 2 (1997), 457–462; B. D. Greenberg et al., "Association between the Serotonin Transporter Promoter Polymorphism and Personality Traits in a Primarily Female Population Sample," *American Journal of Medical Genetics* 96 (2000), 202–216.

57. Eric T. Juengst, "Altering the Human Species? Misplaced Essentialism in Science Policy," in John Rasko, Gabrielle O'Sullivan, and Rachel Ankeny, eds., *The Ethics of Inheritable Genetic Modification* (Cambridge: Cambridge University Press, 2006), 124.

4. Challenges and Risks

1. Greg Bear, "Sisters," in Bear, *Tangents* (London: Orion, 2000), 199–238.

2. "AAAS Report on IGM: Major Findings, Concerns, and Recommendations," in Audrey R. Chapman and Mark S. Frankel, eds., *Designing Our Descendants: The Promises and Perils of Genetic Modifications* (Baltimore: Johns Hopkins University Press, 2003), 349–353.

3. Richard H. Duerr, "A Genome-Wide Association Study Identifies IL23R as an Inflammatory Bowel Disease Gene," *Science* 314 (December 1, 2006), 1461–1463.

4. Antonio Regalado, "Map Quest: New Genetic Tools May Reveal Roots of Everyday Ills—Rapid DNA Tests Can Search Many Variations at Once; Probing Obesity, Memory—One Worry: Statistical Errors," *Wall Street Journal*, April 14, 2006.

5. The most recent count seems to be in the vicinity of 22,500. See Nicholas Wade, "Studies Find Elusive Key to Cell Fate in Embryo," *New York Times*, April 25, 2006. See also Office of Energy, U.S. Department of Energy, "How Many Genes Are in the Human Genome?" Human Genome Project Information, Genomics: GTL, October 27, 2004, available online at http://www.ornl.gov/sci/techresources/Human_Genome/faq/genenumber.shtml.

6. J. C. Venter et al., "The Sequence of the Human Genome," *Science* 291 (February 16, 2001), 1304–1351.

7. B. Modrek and C. Lee, "A Genomic View of Alternative Splicing," *Nature Genetics* 30 (2002), 13–19.

8. P. H. Silverman, "Rethinking Genetic Determinism," *Scientist* 18 (2004), 32–33.

9. Avshalom Caspi et al., "Influence of Life Stress on Depression: Moderation by a Polymorphism in the 5-HTT Gene," *Science* 301 (July 18, 2003), 386–389.

10. For the similar study, see Avshalom Caspi et al., "Role of Genotype in the Cycle of Violence in Maltreated Children," *Science* 297 (August 2, 2002), 851–854.

11. Matt Ridley, *Genome: The Autobiography of a Species in 23 Chapters* (New York: HarperCollins, 1999), 161.

12. For a popular account of Caspi's and related research, see Emily Bazelon, "A Question of Resilience," *New York Times Magazine,* April 30, 2006.

13. Jon W. Gordon, "Genetic Enhancement in Humans," *Science* 283 (March 26, 1999), 2023–2024.

14. For a defense of human genetic engineering as a precautionary strategy in the face of catastrophic environmental risks, see H. Tristram Engelhardt, Jr. and Fabrice Jotterand, "The Precautionary Principle: A Dialectical Reconsideration," *Journal of Medicine and Philosophy* 29/3 (2004), 301–312.

15. Lawrence M. Fisher, "Biochips Signal a Critical Shift for Research and Medicine," *New York Times,* December 21, 1999.

16. Scott D. Patterson and Ruedi H. Aebersold, "Review—Proteomics: The First Decade and Beyond," *Nature Genetics* 33 (2003), 311–323.

17. Mario R. Capecchi, "Human Germline Gene Therapy: How and Why," in Gregory Stock and John Campbell, eds., *Engineering the Human Germline* (New York: Oxford University Press, 2000), 31–48.

18. Enrica Migliaccio et al., "The p66shc Adaptor Protein Controls Oxidative Stress Response and Life Span in Mammals," *Nature* 402 (November 18, 1999), 309–313; Nicholas Wade, "Scientists Link a Single Gene to Longer Life in Mice," *New York Times,* November 18, 1999.

19. S. E. Gabriel et al., "Cystic Fibrosis Heterozygote Resistance to Cholera Toxin in the Cystic Fibrosis Mouse Model," *Science* 266 (October 7, 1994), 107–109.

20. J. I. Rotter and J. Diamond, "What Maintains the Frequencies of Human Genetic Diseases?" *Nature* 329 (1987) 289–290; Jared Diamond, "Curse and Blessing of the Ghetto," *Discover Magazine* 12/3 (March 1991), 60–65. But see B. Spyropoulos et al., "Heterozygote Advantage in Tay-Sachs Carriers?" *American Journal of Human Genetics* 33/3 (May 1981), 375–380; B. Spyropoulos, "Tay-Sachs Carriers and Tuberculosis Resistance," *Nature* 331/6158 (February 25, 1988), 666.

21. Kay Redfield Jamison, *Touched with Fire: Manic Depressive Illness and the Artistic Temperament* (New York: Free Press, 1993).

22. See chapter 1, note 3.

23. The projected cost of the Human Genome Project was $3 billion. The actual cost at completion, for the human genome sequence and a host of related sequence studies, was $2.7 billion. See National Human Genome Research Institute, "The Human Genome Project Completion: Frequently Asked Questions," available online at http://www.genome.gov/11006943.

24. Ridley, *Genome,* 180.

25. For treatments of the Human Genome Diversity Project (HGDP), see Jenny Rear-

don, *Race to the Finish: Identity and Governance in an Age of Genomics* (Princeton, NJ: Princeton University Press, 2005); Henry T. Greely, "Lessons from the HGDP?" *Science* 308 (June 10, 2005), 1554–1555.

26. The Genographic Project is described online at its official Web site: https://www3.nationalgeographic.com/genographic/index.html. See also Benjamin Pimentel, "DNA Study of Human Migration: National Geographic and IBM Investigate Spread of Prehistoric Peoples around World," *San Francisco Chronicle,* April 13, 2005, available online at http://www.sfgate.com/cgi-bin/article .cgi?file=/c/a/2005/04/13/MNGQIC7FT51.DTL. For an account of recent problems experienced by this project, see Amy Harmon, "DNA Gatherers Hit a Snag: The Tribes Don't Trust Them," *New York Times,* December 10, 2006.

27. President's Council on Bioethics, *Beyond Therapy: Biotechnology and the Pursuit of Happiness* (Washington, DC: President's Council on Bioethics, 2003), 165, available online at http://www.bioethics.gov/reports/beyondtherapy/ index.html.

28. Ibid., 182.

29. Ibid., 188.

30. Ibid.

31. Ibid., 189.

32. Ibid., 190.

33. Ibid., 194–195.

34. Ibid., 165. See also Robert W. Fogel, "Changes in the Physiology of Aging during the Twentieth Century," 10, NBER Working Paper No. 11233, March 2005, available online at http://www.nber.org/papers/w11233.

35. Infoplease, "Life Expectancy by Age, 1850–2004," available online at http://www .infoplease.com/ipa/A0005140.html.

36. Rick Lyman, "Census Report Foresees No Crisis over Aging Generation's Health," *New York Times,* March 10, 2006.

37. Nick Bostrom and Toby Ord, "The Reversal Test: Eliminating Status Quo Bias in Applied Ethics," *Ethics* 116 (July 2006), 656–679, available online at http://www .nickbostrom.com/ethics/statusquo.pdf. Bostrom is cofounder of the World Transhumanist Association. Other articles by him and his collaborators can be found on his home page: http://www.nickbostrom.com/. For a discussion of Bostrom's contributions, see Joel Garreau, *Radical Evolution: The Promise and Peril of Enhancing Our Minds, Our Bodies—and What It Means to Be Human* (New York: Doubleday, 2005), 240–248.

38. Sara Goering, "The Ethics of Making the Body Beautiful: What Cosmetic Genetics Can Learn from Cosmetic Surgery," *Philosophy and Public Policy Quarterly* 21/1 (Winter 2001), 21–27.

5. Parents: Guardians or Gardeners?

1. Aldous Huxley, *Brave New World,* 1st Perennial Classics ed. (Harper and Brothers, 1932; reprint, New York: HarperCollins, Harper Perennial, 1998).

2. President's Council on Bioethics, *Beyond Therapy: Biotechnology and the Pursuit of Happiness* (Washington, DC: The President's Council on Bioethics, October 2003), 54–55, available online at http://www.bioethics.gov/reports/beyondtherapy/index.html.

3. Joel Feinberg, "The Child's Right to an Open Future," in Feinberg, *Freedom and Fulfillment: Philosophical Essays* (Princeton, NJ: Princeton University Press, 1992), 76–97.

4. President's Council on Bioethics, *Beyond Therapy: Biotechnology and the Pursuit of Happiness,* 55.

5. Susannah Baruch et al., *Human Germline Genetic Modification: Issues and Options for Policymakers* (Washington, DC: Genetics and Public Policy Center, 2005), 36. The quotation here is from Francis Fukuyama, *Our Posthuman Future: Consequences of the Biotechnology Revolution* (New York: Picador, 2002), 99.

6. This observation is supported by social scientific studies of parent-child attachment. See, for example, John Bowlby, *Childcare and the Growth of Love* (Baltimore, MD: Penguin Books, 1965), chap. 8; Andrea Pound, "Attachment and Maternal Depression," in C. M. Parkes and J. Stevenson-Hinde, eds., *The Place of Attachment in Human Behavior* (London: Tavistock, 1982), 118–130.

7. Marsha Saxton, "Why Members of the Disability Community Oppose Prenatal Diagnosis and Selective Abortion," in Erik Parens and Adrienne Asch, eds., *Prenatal Testing and Disability Rights* (Washington, DC: Georgetown University Press, 2000), 147–164. See also Ann Finger, *Past Due: Disability, Pregnancy, and Birth* (Seattle: Seal Press, 1987); Adrienne Asch and Michelle Fine, "Shared Dreams: A Left Perspective on Disability Rights and Reproductive Rights," in Michelle Fine and Adrienne Asch, eds., *Women and Disabilities: Essays in Psychology, Culture and Politics* (Philadelphia: Temple University Press, 1988), 297–305; Deborah Kaplan, "Prenatal Screening and Its Impact on People with Disabilities," *Fetal Diagnosis and Therapy* 8, supplement 1 (1993), 64–69.

8. Erik Parens and Adrienne Asch, "The Disability Rights Critique of Prenatal Genetic Testing," *Hastings Center Report Special Supplement* 29/5 (1999), S19.

9. Ellen Winner, "The Origins and Ends of Giftedness," *American Psychologist* 55/1 (2000), 159–169.

10. Linda Armstrong Kelly with Joni Rodgers, *No Mountain High Enough: Raising Lance, Raising Me* (New York: Broadway Books, 2005), 149–150, 163.

11. Bill McKibben, *Enough: Staying Human in an Engineered Age* (New York: Henry Holt, 2003), 48–49.

12. Ronald Bailey, *Liberation Biology: The Scientific and Moral Case for the Biotech Revolution* (Amherst, NY: Prometheus Books, 2006), 163.

13. President's Council on Bioethics, *Beyond Therapy,* 93.

14. Harry Adams, "A Human Germline Modification Scale," *Journal of Law, Medicine and Ethics* 32 (2004), 170.

15. Feinberg, "Child's Right to an Open Future"; Dena S. Davis, *Genetic Dilemmas: Reproductive Technology, Parental Choice, and Children's Futures* (New York: Routledge, 2000).

16. Hans Jonas, *Philosophical Essays: From Ancient Creed to Technological Man* (Englewood Cliffs, NJ: Prentice Hall, 1974), 163.

17. William Ruddick, "Parents and Life Prospects," in Onora O'Neill and William Ruddick, eds., *Having Children: Philosophical and Legal Reflections on Parenthood* (New York: Oxford University Press, 1979), 124–137.

18. Ibid., 125.

19. Ibid.

20. Ibid., 128.

21. Daniel Bergner, "The Call," *New York Times Magazine,* January 29, 2006.

22. Feinberg, "Child's Right to an Open Future," 94–95.

23. Ibid., 95.

24. Clive Gillis, "The Restoration of the Roman Catholic Hierarchy in England—The Years of Intrigue and Squirearchy," part 2, in European Institute of Protestant Studies, *The English Churchman,* January 11, 2002, available online at http://www.ianpaisley.org/article.asp?ArtKey=intrigue2. In *The Future of Human Nature* (Cambridge: Polity Press, 2003), the German philosopher Jürgen Habermas attempts to argue that genetic interventions limit freedom more than environmental or educational ones. For a brief presentation and critique of his view, see Nicholas Agar, *Liberal Eugenics: In Defense of Human Enhancement* (Malden, MA: Blackwell, 2004), 116–118.

25. Glenn McGee, *The Perfect Baby: A Pragmatic Approach to Genetics* (Lanham, MD: Rowman and Littlefield, 1997), 127.

26. Adrienne Asch, "Why I Haven't Changed My Mind about Prenatal Diagnosis: Reflections and Refinements," in Parens and Asch, *Prenatal Testing and Disability Rights,* 239.

27. Frances M. Kamm "Is There a Problem with Enhancement?" *American Journal of Bioethics* 5/3 (May–June 2005), 5–14. See also Kamm, "Response to Commentators on "What's Wrong with Enhancement?" *American Journal of Bioethics* 5/3 (2005), W4.

28. Jennifer Egan, "Wanted: A Few Good Sperm," *New York Times Magazine,* March 19, 2006.

29. I owe this reminder to Dena S. Davis. See her *Genetic Dilemmas,* 129–130.

6. Will We Create a "Genobility"?

1. Lee M. Silver, *Remaking Eden* (New York: Avon Books, 1997), 246.

2. Nicholas Wade, *Before the Dawn: Recovering the Lost History of Our Ancestors* (New York: Penguin, 2006), 278–279.

3. Peter Wenz, "Engineering Genetic Injustice," *Bioethics* 19/1 (2005), 1–11.

4. Maxwell Mehlman, *Wondergenes: Genetic Enhancement and the Future of Society* (Bloomington: Indiana University Press, 2003), 116–117.

5. Ibid., 125.

6. Francis Fukuyama, *Our Posthuman Future: Consequences of the Biotechnology Revolution* (New York: Picador, 2002) 9. The Jefferson quotation is from his "Letter to Roger C. Weightman" of June 24, 1826.

7. Ibid., 9–10.

8. Ronald Bailey, *Liberation Biology: The Scientific and Moral Case for the Biotech Revolution* (Amherst, NY: Prometheus Books, 2006), 170.

9. Michael J. Sandel, "The Case against Perfection," *Atlantic Monthly,* April 2004, 51–62. Sandel expands on this argument in his book *The Case against Perfection* (Cambridge: Harvard University Press, 2007), 89–92.

10. Ibid., 62.

11. Quoted in Nicholas Whyte's review of *Beggars in Spain* by Nancy Kress, available online at http://www.nicholaswhyte.info/sf/bis.htm.

12. John Rawls, *A Theory of Justice* (Cambridge: Harvard University Press, 1971).

13. Ibid., 107–108.

14. For an effort to apply Rawls's approach to genetic choice, see Sara Goering, "The Ethics of Enhancement: Gene Therapies and the Pursuit of a Better Human," *Cambridge Quarterly of Healthcare Ethics* 9 (2000), 330–341.

15. Erik Parens, "Genetic Differences and Human Identities: On Why Talking about Behavioral Genetics Is Important and Difficult," *Hastings Center Report Special Supplement* 34/1 (2004), S13.

16. Richard Lynn and Tatu Vanhanen, *IQ and the Wealth of Nations* (Westport, CT: Praeger, 2002). For a discussion of the possible connection between national economic performance and gene enhancement, see Ramez Naam, *More than Human: Embracing the Promise of Biological Enhancement* (New York: Broadway Books, 2005), 56–58.

17. Allen Buchanan, "Genes, Justice and Human Nature," in Allen Buchanan, Dan W.

Brock, Norman Daniels, and Daniel Wikler, *From Chance to Choice: Genetics and Justice* (Cambridge: Cambridge University Press, 2000), 83.

18. The use of religion to justify genetic inequalities is signaled by Frances M. Kamm, "Response to Commentators on 'What's Wrong with Enhancement?'" *American Journal of Bioethics* 5/3 (2005), W4.

19. Nicholas Wade, *Life Script: How the Human Genome Discoveries Will Transform Medicine and Enhance Your Health* (New York: Simon and Schuster, 2001), 177–178.

20. There are many good histories of the eugenics movement. Among them are Daniel J. Kevles, *In the Name of Eugenics* (New York: Alfred A. Knopf, 1985); Martin S. Pernick, *The Black Stork: Eugenics and the Death of "Defective" Babies in American Medicine and Motion Pictures since 1915* (New York: Oxford University Press, 1996); Diane B. Paul, *Controlling Human Heredity: 1865 to the Present* (Amherst, NY: Humanity Books, 1998); Richard Lynn, *Eugenics: A Reassessment* (Westport, CT: Praeger, 2001); Elof Axel Carlson, *The Unfit: A History of a Bad Idea* (Cold Spring Harbor, NY: Cold Spring Harbor Laboratory Press, 2001). For a comprehensive series of resources on eugenics, see the Web site of the Dolan DNA Learning Center at the Cold Spring Harbor Laboratory, available at http://www.eugenicsarchive.org.

21. Philip R. Reilly, *The Surgical Solution: A History of Involuntary Sterilization in the United States* (Baltimore, MD: Johns Hopkins University Press, 1991).

22. Alex Mauron, "Is the Genome the Secular Equivalent of the Soul?" *Science* 291 (February 2, 2001), 831–832.

23. Dorothy Nelkin and M. Susan Lindee, *The DNA Mystique: The Gene as a Cultural Icon* (New York: W. H. Freeman, 1995).

24. Paul, *Controlling Human Heredity*, 42.

25. Dolan DNA Learning Center, Cold Spring Harbor Laboratory, EugenicsArchive.org: Themes: "Talent," available online at http://www.eugenicsarchive.org/html/eugenics/static/themes/7.html.

26. Nancy L. Gallagher, *Breeding Better Vermonters: The Eugenics Project in the Green Mountain State* (Hanover, NH: University Press of New England, 1999).

27. Paul, *Controlling Human Heredity*, 135.

28. For an analysis of the features that made the eugenics movement so harmful, see Allen Buchanan, Dan W. Brock, Norman Daniels, and Daniel Wikler, *From Chance to Choice: Genetics and Justice*, chap. 2. Buchanan et al. offer a more positive vision of the prospects of eugenics in our era, as does Philip Kitcher in his *Lives to Come: The Genetic Revolution and Human Possibilities* (New York: Simon and Schuster, 1996).

29. In 1942 the U.S. Supreme Court unanimously overturned a sterilization law in an opinion that termed procreation "one of the basic civil rights of man": *Skinner v. Oklahoma*, 316 U.S. 535 (1942). The court further expanded the scope of privacy and reproductive freedom in *Griswold v. Connecticut* (1965), *Eisenstadt v. Baird*

(1972), and *Roe v. Wade* (1973). The last is currently imperiled, raising the question of whether the tide of opinion could be reversed. For a review of the judicial history of procreative liberty in the U.S., see Judith F. Daar, *Reproductive Technologies and the Law* (Dayton, OH: LexisNexis, 2005).

30. Troy Duster, *Backdoor to Eugenics,* 2nd ed. (New York: Routledge, 2003).

31. Carl Elliott, *Better than Well: American Medicine Meets the American Dream* (New York: W. W. Norton, 2003), 248.

32. Diane B. Paul, "Is Human Genetics Disguised Eugenics?" in Robert F. Weir et al., eds., *Genes and Human Self-Knowledge: Historical and Philosophical Reflections on Modern Genetics* (Iowa City: University of Iowa Press, 1994), 67–83.

33. Robert Wright, "Achilles' Helix," *New Republic,* July 9 and 16, 1990, 27. See also David King, "Eugenic Tendencies in Modern Genetics," in Peter Doherty and Agneta Sutton, eds., *Man-Made Man: Ethical and Legal Challenges in Genetics* (Dublin, Ireland: Four Courts Press, 1997), 71–82.

34. C. S. Lewis, *The Abolition of Man* (New York: Macmillan, 1947), 36–37.

35. G. Jasienska et al., "Large Breasts and Narrow Waists Indicate High Reproductive Potential in Women," *Proceedings of the Biological Sciences* 271 (2004), 1213–1237.

36. Alan F. Dixson et al., "Masculine Somatotype and Hirsuteness as Determinants of Sexual Attractiveness to Women," *Archives of Sexual Behavior* 32/1 (2003), 29–39; John S. Gillis, *Too Tall, Too Small* (Champaign, IL: Institute for Personality and Ability Testing, 1982), chap. 2.

37. In a different direction, Simon Baron-Cohen, director of the Autism Research Centre at Cambridge University, has argued that assortative mating between highly driven personality types that he calls "systemizers," who are likely to occur in an environment of demanding technical academic programs, may be a factor in the apparently increasing incidence of autism. See "The Assortative Mating Theory: A Talk with Simon Baron-Cohen," *Edge: The Third Culture,* available online at http://www.edge.org/3rd_culture/baron-cohen05/baron-cohen05_index.html.

38. David Plotz, *The Genius Factory: The Curious History of the Nobel Prize Sperm Bank* (New York: Random House, 2005).

39. David Plotz, "The 'Genius Babies' and How They Grew," *Slate,* February 8, 2001, available online at http://slate.com/id/100331/.

40. Quoted in Richard Barlow, "Hoping for Extra-Brainy Offspring, Some Seek Eggs of Ivy Women," *Valley News* (West Lebanon, NH), January 24, 1999.

41. Mark S. Frankel and Audrey R. Chapman, *Human Inheritable Genetic Modifications: Assessing Scientific, Ethical, Religious, and Policy Issues* (Washington, DC: American Association for the Advancement of Science, 2000), available online at http://www.aaas.org/spp/sfrl/projects/germline/report.pdf.

42. Lewis Thomas, "The Technology of Medicine," *New England Journal of Medicine* 285 (1971), 1367.

43. Ibid., 1368.

44. Donald G. McNeil, Jr., "In Raising the World's I.Q., the Secret's in the Salt," *New York Times*, December 15, 2006.

45. Buchanan, "Genes, Justice and Human Nature," 78.

7. Playing God

1. President's Commission for the Study of Ethical Problems in Medicine and Biomedical and Behavioral Research, *Splicing Life: A Report on the Social and Ethical Issues of Genetic Engineering with Human Beings* (Washington, DC: President's Commission for the Study of Ethical Problems in Medicine and Biomedical and Behavioral Research, 1982).

2. Paul Ramsey, *Fabricated Man: The Ethics of Genetic Control* (New Haven: Yale University Press, 1970), 138.

3. Leon Kass, *The Beginning of Wisdom* (New York: Free Press, 2003), 217–243.

4. Phillip Elmer-Dewitt, "The Genetic Revolution," *Time,* January 17, 1994, 46.

5. Poll conducted October 29 and November 2, 1997. The results of this poll are discussed by Lee M. Silver in *Challenging Nature* (New York: HarperCollins, 2006), 324f. In a 2004 poll of Americans' attitudes toward reproductive testing, over a third of those polled listed "playing God" as what most worries them in connection with human control of reproduction, while another third worried most about its use for the wrong purposes—A. Kalfoglou, K. Suthers, J. Scott, and K. Hudson, *Reproductive Genetic Testing: What America Thinks* (Washington, DC: Genetics and Public Policy Center, 2004), available online at http://www.dnapolicy.org/images/reportpdfs/ReproGenTestAmericaThinks.pdf.

6. Pope John Paul II, "Biological Research and Human Dignity," *Origins* 12 (November 4, 1982), 343.

7. Pope John Paul II, "The Ethics of Genetic Manipulation: Address to the World Medical Association," *Origins* 13 (November 17, 1983), 388.

8. International Theological Commission, "Communion and Stewardship: Human Persons Created in the Image of God" (2004), §90, available online at http://www.vatican.va/roman_curia/congregations/cfaith/cti_documents/rc_con_cfaith_doc_20040723_communion-stewardship_en.html.

9. Pope John Paul II, "Ethics of Genetic Manipulation," 388.

10. Ibid. The commission under Cardinal Ratzinger that approved germline therapies also declared that "changing the genetic identity of man as a human person

through the production of an infrahuman being is radically immoral." International Theological Commission, "Communion and Stewardship," §91.

11. Jonathan Glover, *What Sort of People Should There Be?* (Harmondsworth, England: Penguin, 1984), 46.

12. Ted Peters, *Playing God: Genetic Determinism and Human Freedom* (New York: Routledge, 2003), 14-15. In this discussion, Peters also attributes some of these views to the Protestant theologian Ronald Cole-Turner. The Bouma quotation comes from *Christian Faith, Health, and Medical Practice* (1989).

13. Quoted in Laurie Zoloth, "Uncountable as the Stars: Inheritable Genetic Intervention and the Human Future—A Jewish Perspective," in Audrey R. Chapman and Mark S. Frankel, eds., *Designing Our Descendants: The Promises and Perils of Genetic Modifications* (Baltimore: Johns Hopkins University Press, 2003), 224.

14. Donald G. McNeil, Jr., "Circumcision's Anti-AIDS Effect Found Greater Than First Thought," *New York Times,* February 23, 2007. For the initial news reports on these studies, see Sharon LaFraniere, "Circumcision Studied in Africa as AIDS Preventive," *New York Times,* April 28, 2006, and Donald G. McNeil, Jr., "Circumcision Reduces Risk of AIDS, Study Finds," *New York Times,* December 13, 2006.

15. Liebe F. Cavalieri, quoted in President's Commission for the Study of Ethical Problems, *Splicing Life,* 62. See, too, Edward O. Wilson, *Consilience* (New York: Alfred A. Knopf, 1998), 277.

16. Emmanuel Agius, "Germ-Line Cells: Our Responsibilities for Future Generations," in Salvino Busuttil, Emmanuel Agius, Peter Serracino Inglott, and Tony Macelli, eds., *Our Responsibilities towards Future Generations* (Valletta, Malta: Foundation for International Studies, 1990), 133-143.

17. George J. Annas, Lori B. Andrews, and Rosario M. Isasi, "Protecting the Endangered Human: Toward an International Treaty Prohibiting Cloning and Inheritable Alterations," *American Journal of Law and Medicine* 28 (2002), 151-178.

18. Luis Archer, "Genetic Testing and Gene Therapy: The Scientific and Ethical Background," in Peter Doherty and Agneta Sutton, eds., *Man-Made Man: Ethical and Legal Challenges in Genetics* (Dublin, Ireland: Four Courts Press, 1997), 42.

19. President's Council on Bioethics, "Sixth Meeting Friday, September 13, 2002," Session 7: Enhancement 5: Genetic Enhancement of Muscle, available online at http://www.bioethics.gov/transcripts/sep02/session7.html.

20. For extensive treatments of evolution and aging, see Michael Rose, *Evolutionary Biology of Aging* (New York: Oxford University Press, 1991); and Rose, *The Long Tomorrow: How Advances in Evolutionary Biology Can Help Us Postpone Aging* (New York: Oxford University Press, 2005). For a view that presents an active role for genes in determining longevity, see Adam Antebi, "The Tick-Tock of Aging?" *Science* 310 (December 23, 2005), 1911-1913.

21. Enrica Migliaccio et al., "The p66shc Adaptor Protein Controls Oxidative Stress Response and Life Span in Mammals," *Nature* 402 (November 18, 1999), 309–313; Nicholas Wade, "Scientists Link a Single Gene to Longer Life in Mice," *New York Times,* November 18, 1999.

22. Daniel C. Dennett, "Show Me the Science," *New York Times,* August 28, 2005.

23. Alan Herbert et al., "A Common Genetic Variant Is Associated with Adult and Childhood Obesity," *Science* 312 (April 14, 2006), 279–283.

24. Marc Santora, "Concern Grows over Increase in Diabetes around World," *New York Times,* June 11, 2006.

25. Glover, *What Sort of People Should There Be?* 182–185.

26. For a good overview of behavioral and cognitive genetic research, see Dean Hamer and Peter Copeland, *Living with Our Genes: Why They Matter More than You Think* (New York: Doubleday, 1998). Eric Parens, Audrey Chapman, and Nancy Press discuss some of the limitations of this research in their *Wrestling with Behavioral Genetics* (Baltimore, MD: Johns Hopkins University Press, 2006).

27. Arthur R. Peacocke, *God and the New Biology* (San Francisco: Harper and Row, 1986), 121.

28. Nicholas Wade, "Nice Rats, Nasty Rats: Maybe It's All in the Genes," *New York Times,* July 25, 2006.

29. Octavia Butler, *Dawn* (New York: Warner Books, 1987), 39.

30. Robert Graves and Raphael Patai, *Hebrew Myths: The Book of Genesis* (New York: Doubleday, 1964), 65–69.

31. Butler, *Dawn,* 245.

32. Ibid., 247.

33. Ibid.

34. Ibid., 248.

35. Octavia Butler, *Adulthood Rites* (New York: Warner Books, 1987), 233–234.

36. Octavia Butler, *Imago* (New York: Popular Library, 1989), 220.

37. The Universal Declaration on the Human Genome and Human Rights, adopted unanimously and by acclamation by the UNESCO General Conference at its twenty-ninth session on November 11, 1997, Article 1, available online at http://portal.unesco.org/shs/en/ev.php-URL_ID=1881&URL_DO=DO_TOPIC&URL_SECTION=201.html.

38. Leon Kass, *Toward a More Natural Science: Biology and Human Affairs* (New York: Free Press, 1985), 272–273.

39. Butler, *Adulthood Rites,* 80.

40. The same ambivalence about extraterrestrials' domination/protection of human beings is evidenced in Butler's landmark story "Bloodchild" (1984). For a discussion of these themes in Butler, see Elyce Rae Helford, "'Would You Really Rather

Die than Bear My Young?' The Construction of Gender, Race, and Species in Octavia E. Butler's 'Bloodchild,'" *African American Review* 28 (1994), 259–271.

41. Butler, *Dawn*, 154.

42. Roger Lincoln Shinn, *The New Genetics: Challenges for Science, Faith and Politics* (Wakefield, RI: Moyer Bell, 1996), 125.

43. Peters, *Playing God*, 39.

44. Peters, *Playing God*, 205. See also Arthur R. Peacocke, *Creation and the World of Science*, Bampton Lectures, 1978 (Oxford: Oxford University Press, Clarendon Press, 1979), 301–315.

8. The Choices Ahead

1. Rachel Shabi, "Baby Chase," *Guardian Unlimited*, June 26, 2004, available online at http://www.guardian.co.uk/weekend/story/0,3605,1246289,00.html#article_continue. Shabi misreports the name of the family as the Mastersons.

2. "Couple Angered by Baby Ruling," BBC Newsnight Scotland, October 18, 2000, available online at http://news.bbc.co.uk/2/hi/uk_news/scotland/977450.stm.

3. Shabi, "Baby Chase."

4. Human Fertilisation and Embryology Authority, *Sex Selection: Options for Regulation: A Report on the HFEA's 2002–2003 Review of Sex Selection Including a Discussion of Legislative and Regulatory Options*, available online at https://centres.hfea.gov.uk/AboutHFEA/Consultations/Final%20sex%20selection%20main%20report.pdf. For a discussion of preconception sex selection using sperm-sorting techniques, see John Robertson, "Preconception Gender Selection," *American Journal of Bioethics* 1/1 (January 2001), 2–9, and the accompanying peer commentaries by Norman Daniels, Rebecca Dresser, and others, 10–39.

5. Amartya Sen, "More than 100 Million Women Are Missing," *New York Review of Books* 37/20 (December 20, 1990), 61–66. Sen's numbers and explanations have since been contested. See Ansley J. Coale, "Excess Female Mortality and the Balance of the Sexes in the Population: An Estimate of the Number of 'Missing Females,'" *Population and Development Review* 17/3 (September 1991), 517–523; Robert J. Barro, "Economic Viewpoint: The Case Of Asia's 'Missing Women,'" *Business Week*, February 28, 2005, available online at http://post.economics.harvard.edu/faculty/barro/bw/bw05_02_28.pdf.

6. This is not to say that there are no risks for the donor child. For a discussion of these, see Susan M. Wolf, Jeffrey P. Kahn, and John E. Wagner, "Using Preimplantation Genetic Diagnosis to Create a Stem Cell Donor: Issues, Guidelines and Limits," *Journal of Law, Medicine and Ethics* 31/3 (2003), 327–339. The risks are also given fictional form in Jodi Picoult's novel *My Sister's Keeper* (New York: Atria Books, 2004).

7. On sex preferences in developed countries, see Edgar Dahl, "Procreative Liberty: The Case for Preconception Sex Selection," *Reproductive BioMedicine Online* 7/4 (2003), 380–384.

8. Birth order, sibling interval, and sibling crowding have all been implicated in IQ performance and life achievement. See Robert Zajonc, "Family Configuration and Intelligence," *Science* 192 (April 16, 1976), 227–236; Peter Lindert, "Sibling Position and Achievement," *Journal of Human Resources* 12/2 (1977), 220–241; Dalton Conley, *Pecking Order: Which Siblings Succeed and Why* (New York: Pantheon Books, 2004).

9. Bioethicists advancing the argument include Tabitha Powledge, "Unnatural Selection: On Choosing Children's Sex," in Helen B. Holmes, Betty B. Hoskins, and Michael Gross, eds. *The Custom-Made Child? Woman-Centered Perspectives* (Clifton, NJ: Humana, 1981), 197; Dorothy Wertz and John C. Fletcher, "Fatal Knowledge? Prenatal Diagnosis and Sex Selection," *Hastings Center Report* 19/1 (1989), 21–27.

10. Michael D. Bayles, *Reproductive Ethics* (Englewood Cliffs, NJ: Prentice Hall, 1984), 34–36.

11. Leslie Laurence, "The Bargain Baby-Maker," *New York Magazine,* February 9, 1998, available online at http://nymag.com/nymetro/health/features/2163/index .html.

12. Susannah Baruch, David Kaufman, and Kathy L. Hudson, "Genetic Testing of Embryos: Practices and Perspectives of U.S. IVF Clinics," *Fertility and Sterility,* published online on September 21, 2006.

13. Dr. Sandra A. Carson, Baylor College of Medicine, in a telephone interview with the author, May 23, 2006.

14. Isaac Rabino, "The Impact of Activist Pressures on Recombinant DNA Research," *Science, Technology and Human Values* 16/1 (1991), 70–87. See also Paul S. Naik, "Biotechnology through the Eyes of an Opponent: The Resistance of Activist Jeremy Rifkin," *Virginia Journal of Law and Technology* 5 (Spring 2000), §55, available online at http://www.vjolt.net/vol5/issue2/v5i2a5-Naik.html#text92.

15. Francis Fukuyama, *Our Posthuman Future: Consequences of the Biotechnology Revolution* (New York: Picador, 2002), 10.

16. Francis Fukuyama and Franco Furger, *Beyond Bioethics: A Proposal for Modernizing the Regulation of Human Biotechnologies* (Washington, DC: Paul H. Nitze School of Advanced International Studies, 2006).

17. This problem is developed in Ramez Naam, *More than Human: Embracing the Promise of Biological Enhancement* (New York: Broadway Books, 2005), 70.

18. Mary Ann Warren, *Gendercide: The Implications of Sex Selection* (Totowa, NJ: Rowman and Allanheld, 1985), 187, 185.

19. Maxwell Mehlman and Kirsten M. Rabe, "Any DNA to Declare: Regulating Off-

shore Access to Genetic Enhancement?" *American Journal of Law and Medicine* 28 (2002), 179-213.

20. Julian Savulescu, "In Defense of Selection of Nondisease Genes," *American Journal of Bioethics* 1/1 (2001), 18.

21. For reviews of current regulatory and legal options, see Maxwell Mehlman, "How Will We Regulate Genetic Enhancement?" *Wake Forest Law Review* 34 (1999), 671-714; Mehlman and Rabe, "Any DNA to Declare"; N. King, "RAC Oversight of Gene Transfer Research: A Model Worth Extending"; and Rebecca Dresser, "Gene Modification of Preimplantation Embryos: Toward Adequate Research Policies," *Milbank Quarterly* 82/1 (2004), 195-214.

22. Dresser, "Gene Modification of Preimplantation Embryos."

23. Susannah Baruch et al., *Human Germline Genetic Modification: Issues and Options for Policymakers* (Washington, DC: Genetics and Public Policy Center, 2005).

24. J. Cohen et al., "Birth of Infant after Transfer of Anucleate Donor Oocyte Cytoplasm into Recipient Eggs," *Lancet* 350 (1997), 186-187.

25. Center for Genetics and Society, "Ooplasmic Transfer," available online at http://genetics-and-society.org/analysis/promodeveloping/ooplasmic.html.

26. U.S. Food and Drug Administration (FDA), "Letter to Sponsors/Researchers: Human Cells Used in Therapy Involving the Transfer of Genetic Material by Means Other Than the Union of Gamete Nuclei," July 6, 2001, available online at http://www.fda.gov/cber/ltr/cytotrans070601.htm.

27. Eve E. Slater, "Today's FDA," *New England Journal of Medicine* 352 (January 20, 2005), 293-297.

28. For discussions of the strengths and weaknesses of the decentralized U.S. system of research oversight, see John A. Robertson, "Ten Ways to Improve IRBs," *Hastings Center Report* 9/1 (1979), 29-33; and Lucas Bergkamp, "American IRBs and the Dutch Research Ethics Committees: How They Compare," *IRB* 10/6 (September-October 1988), 5.

29. Mehlman and Rabe, "Any DNA to Declare"; Kathy L. Hudson, "Preimplantation Genetic Diagnosis: Public Policy and Public Attitudes," *Fertility and Sterility* 85/6 (June 2006), 1638-1645.

30. See, for example, Mehlman's discussion of the American Medical Association's poorly developed 1994 policy statement on genetic enhancement: Mehlman, "How Will We Regulate Genetic Enhancement," 693-694.

31. Marlise Simons, "French Uproar over Right to Death for Unborn," *New York Times,* October 19, 2001; A. Carmi, "Wrongful Life: An Israeli Case," *Medicine and Law* 9/2 (1990), 777-781; Margery W. Shaw, "To Be or Not to Be? That Is the Question," *American Journal of Human Genetics* 36 (1984), 1-9; Julie A. Greenberg, "Reconceptualizing Preconception Torts," *Tennessee Law Review* 64 (1997), 315-357; N. Pri-

aulx, "Joy to the World! A (Healthy) Child Is Born! Reconceptualizing 'Harm' in Wrongful Conception," *Social and Legal Studies* 13/1 (2004), 5–27.

32. *Berman v. Allan,* 80 N.J. 421, 404 A. 2d 8 (1979).

33. Cynthia B. Cohen, "'Give Me Children or I Shall Die!' New Reproductive Technologies and Harm to Children," *Hastings Center Report* 26/2 (1996), 23; Ronald M. Green, "Parental Autonomy and the Obligation Not to Genetically Harm One's Child: Implications for Clinical Genetics," *Journal of Law, Medicine and Ethics* 25/1 (1997), 5–15.

34. George J. Annas, "The Patient's Right to Safety—Improving the Quality of Care through Litigation against Hospitals," *New England Journal of Medicine* 354/19 (May 11, 2006), 2063–2066.

35. Shaw, "To Be or Not to Be?" 9.

36. Sondra Wheeler, "Parental Liberty and the Right of Access to Germ-Line Intervention: A Theological Appraisal of Parental Power," in Audrey R. Chapman and Mark S. Frankel, eds., *Designing Our Descendants: The Promises and Perils of Genetic Modifications* (Baltimore: Johns Hopkins University Press, 2003), 246.

37. William Saletan, "In Your Eye: If Steroids Are Cheating, Why Isn't LASIK?" *Slate,* April 18, 2005, available online at http://www.slate.com/id/2116858/. It is reported that LASIK eye surgery has increased exponentially at the U.S. military academies and in the armed forces. See David S. Cloud, "Perfect Vision, via Surgery, Is Helping and Hurting Navy," *New York Times,* June 20, 2006.

38. Dan Brock and Norman Daniels, "Why Not the Best?" in Allen Buchanan, Dan W. Brock, Norman Daniels, and Daniel Wikler, *From Chance to Choice: Genetics and Justice* (Cambridge: Cambridge University Press, 2000), 174.

39. Leon R. Kass, "Babies by Means of In Vitro Fertilization: Unethical Experiments on the Unborn," *New England Journal of Medicine* 285 (1971), 1174–1179; Paul Ramsey, "Shall We 'Reproduce'?" part I, "The Medical Ethics of In Vitro Fertilization," and part II, "Rejoinders and Future Forecasts," *JAMA: The Journal of the American Medical Association* 220/10 (1974), 1346–1350; 220/11 (1974), 1480–1485.

40. John A. Robertson, *Children of Choice* (Princeton, NJ: Princeton University Press, 1994), 75–76; Robertson, "Liberty, Identity, and Human Cloning," *Texas Law Review* 76/6 (May 1998), 1405.

41. Christopher Carlson, "On Average, How Many Mutations Take Place Every Minute?" MadSci Network, January 22, 2001, available online at http://www.madsci.org/posts/archives/jan2001/980213843.Ge.r.html. Henry Harpending and Gregory Cochran state that the highest known spontaneous mutation rates for rare genetic disorders are on the order of one in ten thousand. See their "Genetic Diversity and Genetic Burden in Humans," *Infection, Genetics and Evolution* 6 (2006), 154–162.

42. Carlson, "On Average, How Many Mutations?"

43. Lewis Thomas, "The Wonderful Mistake," in Thomas, *The Madonna and the Snail: More Notes from a Biology Watcher* (New York: Viking, 1979), 27–30.

44. Lee M. Silver offers a hypothetical case involving such a couple in an argument on behalf of human genetic engineering. See his *Challenging Nature* (New York: HarperCollins, 2006), 342–344.

45. S. M. Elbashir, "Duplexes of 21-Nucleotide RNAs Mediate RNA Interference in Cultured Mammalian Cells," *Nature* 411 (2001), 494–498.

46. Rebecca Dresser, "Designing Babies: Human Research Issues," *IRB: Ethics and Human Research* 26/5 (2004), 1–8.

47. Garrett Hardin, "The Tragedy of the Commons," *Science* 162 (December 13, 1968), 1243–1248.

48. Julian Savulescu, "New Breeds of Humans: The Moral Obligation to Enhance," *RBMOnline* 10, supplement 1 (2005), 39, available online at http://www.rbmonline.com/4DCGI/Article/Detail?38%091%09=%201643%09.

49. Arthur Caplan, "Is Better Best?" *Scientific American,* September 2003, 104–105.

50. Dean H. Hamer et al., "A Linkage between DNA Marker on the X Chromosome and Male Sexual Orientation," *Science* 261 (July 16, 1993), 321–327; Mark Schoofs, "Gene–ocide: Can Scientists 'Cure' Homosexuality by Altering DNA?" *Village Voice,* July 1, 1997, 40ff.

51. Jonathan Tolins, *The Twilight of the Golds: A Play in Two Acts* (New York: S. French, 1994). See also Schoofs, "Gene–ocide," 40.

52. Albert Mohler, "Is Your Baby Gay? What If You Could Know? What If You Could Do Something about It?" www.albertmohler.com, March 2, 2007, available online at http://www.albertmohler.com/blog_read.php?id=891.

53. Hamer made this remark in a lecture in a Summer Faculty Workshop on the Ethical, Legal and Social Implications of the Human Genome Project, Howard University, June 16, 2005.

54. Darshak M. Sanghavi, "Essay: Wanting Babies Like Themselves, Some Parents Choose Genetic Defects," *New York Times,* December 5, 2006. See also Liza Mundy, "A World of Their Own," *Washington Post Magazine,* March 27, 2002.

55. Laura Blackburn, "U.K. Embryos May Be Screened for Cancer Risk," *Science* 312 (May 19, 2006), 984.

56. Ibid.

57. Ronald Cole-Turner, "Do Means Matter? Evaluating Technologies of Human Enhancement," *Report from the Institute for Philosophy and Public Policy* 18 (Fall 1998), available online at http://www.publicpolicy.umd.edu/IPPP/fall98/do_means_matter.htm.

58. Edward O. Wilson, *Consilience* (New York: Alfred A. Knopf, 1998), 270, 277.

Glossary

5-HTT A gene that controls various brain functions. The short form (allele) of this gene is associated with depressive symptoms and diagnosable depression.

AAAS American Academy for the Advancement of Science.

ACTN3 A gene that apparently gives rise to a higher proportion of the fast-twitch muscle fibers used in high-speed, short-distance sports events.

Allele An alternative form of a gene located at a specific position on a specific chromosome. In humans and other organisms having two copies of each chromosome, each gene exists in at least two versions, or alleles.

Alternative splicing The ability of a single gene to give rise to multiple transcripts of its DNA and thus multiple distinct proteins with multiple functions.

ASPM A gene that regulates brain size.

Blastomere One of the cells that make up the early embryo.

BRCA1, BRCA2 Genes with mutations that greatly increase an individual's risk of breast, ovarian, or prostate cancer.

CCR5 A gene that codes for the receptor that is the docking point for the AIDS virus.

Chimera	An animal (usually a mouse or rat in genetic research) in which cells from other animals (or other species of animals) have been inserted early in embryonic development.
Chromosomes	Units of genetic material, composed of DNA and protein structures, across which the three billion pairs of nucleotide letters of the genome are distributed. The human genome has forty-six chromosomes.
CRE recombinase	An enzyme that, when activated by a special drug, deletes or inverts all DNA letters that it finds between two DNA sequences called LoxP sites.
Cystic fibrosis	A genetic disorder in which gene mutations impair the ability of cell membranes to transport salt (chloride). This can cause a life-threatening thickening and buildup of mucus in the lungs.
DCDC2	A gene active in the reading centers of the human brain. Large deletions in a regulatory region of the gene have been found in one of every five dyslexics tested.
DNA	Deoxyribonucleic acid. The long molecule composed of nucleotides that constitutes the molecular basis of heredity. The nucleotides are lined up on two connecting complementary strands that twist along their length to form a double-helix structure.
Dominant gene or trait	A gene or trait that hides a recessive trait. Usually having one allele causing a dominant trait is sufficient for the trait to manifest itself. *See also* Recessive gene or trait.
EPO	Erythropoietin, a hormone responsible for stimulating red blood cell production.
Fanconi anemia	A rare genetic disease that affects children and causes characteristic short stature, skeletal anomalies, increased incidence of solid tumors and leukemias, and bone marrow failure.
FOXP2	A gene associated with speech acquisition.
Gene	A sequence of DNA letters (nucleotides) that codes for the proteins and other chemical elements that make up the

structures and direct the functions of a biological organism. Genes consist of a long strand of DNA containing a coding sequence, which determines what the gene produces, and a promoter (or regulatory) region, which controls how much of the gene product is produced. These sequences can be interrupted by long stretches of non-coding DNA.

Gene chip
A silicon microarray engraved with sequences of DNA used to test for the expression of DNA sequences in cells.

Gene doping
The non-therapeutic use of cells, genes, or genetic elements, or the modulation of gene expression to improve athletic performance.

Gene surgery
The use of targeted gene transfer techniques to repair or eliminate disease-causing DNA sequences. *See also* Gene therapy.

Gene therapy
A technique for correcting defective genes that are responsible for disease development. Some prefer the more general term "gene transfer" because there have been few therapeutic results of this research and because the technology can be used for non-therapeutic, enhancement purposes. *See also* Germline gene therapy; Somatic cell gene therapy.

Genome
The total complement of genes in an organism.

Genotype
The genetic makeup encoded in an individual's DNA. The term is often used in contrast with "phenotype" to distinguish genetic from environmental factors shaping the organism.

Germline gene therapy (or germline gene transfer)
A technique for altering genes in the sex cells (or cells of the early embryo), with the result that the changes are capable of being passed on to subsequent generations.

HACs
Human artificial chromosomes.

Heritability
The percentage of observed variation among individuals that is attributable to genes as opposed to environment. This is not the same thing as "hereditary." Many features are highly inherited—they are much more the product of

ancestry than they are of environment—but they have low heritability because genes cause very little variance from individual to individual. Having two legs is an inherited feature of human beings, but it is not highly heritable, since most of the variation between people having fewer than two legs is environmentally caused.

Heterozygote

An organism that has one copy of a gene variant (allele); a heterozygous genotype.

Heterozygote advantage

This occurs when an individual having one copy of a gene variant (allele) has a relatively higher reproductive/survival rate than one with either no or two copies of the allele.

HFEA

Human Fertilisation and Embryology Authority, the agency responsible for oversight of assisted reproductive technologies in Britain.

hGH

Human growth hormone.

HLA

Human leukocyte antigen system, the complex of genes controlling immune resistance.

Homologous recombination

Utilization of the cell's own gene repair mechanisms to make site-specific gene targeting and gene alteration possible.

Homologue

A similar (but not necessarily identical) gene carried in different species that indicates a common ancestry.

Homozygote

An organism that has two copies of the same gene variant (allele); homozygous genotype.

Human Genome Project (HGP)

The thirteen-year-long project (from 1990 to 2003) coordinated by the U.S. Department of Energy and the National Institutes of Health to sequence the three billion nucleotides contained in the human genome.

Huntington's disease

A dominant inherited disorder that causes serious neurological degeneration later in life. Sometimes called Huntington's chorea.

Hutchinson-Gilford syndrome

A genetic disease that causes rapid aging in children.

Hydroxyurea	A drug approved by the FDA to treat certain types of leukemia and other cancers. It has been found to reduce the severity of sickle cell disease in many patients.
Hypobaric training	Athletic training regimens at low air pressure / high altitude.
IGF-I	Insulinlike growth factor, a chemical agent in the body that multiplies the number of cells responsible for the growth of muscle tissue. Small variations in the canine version of the gene producing IGF-I have also been found to be responsible for variations in size among breeds of dogs.
IL2RG	A gene that codes for the proteins that make up the interleukin-2 receptor. These receptors are needed for the proper functioning of key cells of the immune system.
INDY	A gene (the letters stand for "I'm not dead yet") that controls the movement of nutrients into cells and whose modification is associated with increased longevity.
Insertional mutagenesis	The accidental insertion of a DNA sequence in the wrong place on the genome. This can cause cancer or other serious illnesses.
ISS	Idiopathic short stature; shortness whose cause (genetic or otherwise) is unknown.
IVF	In vitro fertilization, a technique with which egg cells are fertilized by sperm outside the woman's womb.
IVM	In vitro maturation, a still-undeveloped technology for maturing eggs (oocytes) outside the woman's body to the point where they are ready for fertilization.
Junk DNA	Sequences of DNA in the genome that do not code for proteins or other regulatory products. It is not established, however, that all such DNA has no function.
Knockin (or knockout) mice	Mice in which specific DNA sequences have been inserted (or removed). These mice are test beds for gene function and dysfunction.
Lesch-Nyhan Syndrome	A rare X-chromosone-linked recessive disorder that causes a buildup of uric acid in all body fluids and leads from birth

on to such symptoms as severe gout, poor muscle control, and moderate retardation. A striking feature of the syndrome is self-mutilating lip and finger biting that begins in the second year of life.

LoxP sites DNA sequences thirty-four nucleotides in length that can be used to remove or invert specific DNA sequences located between them.

Mendelian traits Genetic traits or diseases caused by a variation or mutation in a single gene and usually inherited in dominant or recessive patterns. *See also* Polygenic traits.

Microarrays Silicon wafers engraved with sequences of DNA used to test for the expression of DNA sequences in cells. *See also* Gene chip.

Microcephalin A gene controlling the number of neurons produced during fetal development.

Mutations Changes to the nucleotide sequence of genetic material.

Myostatin A naturally occurring protein that seems to limit muscle growth throughout embryonic development and adult life.

Neurons Cells in the nervous system that function to process and transmit information.

NIH National Institutes of Health.

Nucleotides The four subunits (bases, or chemical letters) of a DNA strand: A (adenine), C (cytosine), G (guanine), and T (thymine). Three billion of these letters make up the DNA sequence of the human genome. One strand of these letters pairs up with its complement (As with Ts; Cs with Gs) to create a double helix of "base pairs" that allows DNA copying when cells divide.

Original position A standpoint posited by the Harvard philosopher John Rawls, who maintains that members of a society should select the basic principles of justice governing their long-term relationships by assuming that all members of the society are equal and lack knowledge of their special privileges or liabilities. *See also* Veil of ignorance.

PCR Polymerase chain reaction, a technology used to synthesize large numbers of a particular sequence of DNA.

PDYN Prodynorphin, a protein that is a precursor to a number of endorphins (opiatelike molecules involved in learning, the experience of pain, and social attachment). Chimpanzees and humans both have the gene that produces PDYN, but it is expressed differently in the two species.

PGD Preimplantation genetic diagnosis.

Phenotype The visible or detectable characteristics of an organism shaped by the organism's genes and environment.

PLAAP "Parental Love Almost Always Prevails," a proposed psychological principle explaining why parents' advance selection of children's genetic traits may not interfere with full parental bonding to any resulting children.

Pleiotropy The contribution of a single gene to more than one trait.

Polygenic traits Complex traits caused by two or more genes or gene components working together. The opposite of Mendelian traits.

Positional advantage What people seek, or secure, when they try to do better than others in competitive activities.

Progerias Genetic diseases that cause accelerated aging.

Promoter A sequence of DNA that forms part of a gene and that is located upstream of the coding region of the gene. It activates or represses the gene and controls the expression of the gene's products. *See also* Regulatory region.

Proteomics The study of proteins, particularly their structures and functions, as well as the full complement of proteins produced in a living organism.

Punnett Square A rectangle subdivided into four equal compartments, used by geneticists to calculate the genetic inheritance patterns of Mendelian traits.

RAC The NIH's Recombinant DNA Advisory Committee.

Recessive gene or trait   A recessive gene or trait refers to an allele that causes a phenotype only when both chromosomes carry that allele. *See also* Dominant gene or trait.

Regulatory region   A sequence of DNA that activates or represses a gene and controls the expression of the gene's products. *See also* Promoter.

Reprogenetics   The merging of reproductive and genetic technologies.

Robo1   A gene that guides connections during fetal development between the brain's two hemispheres.

Savior child   A baby conceived to provide a tissue match for an existing sick sibling.

Sex-linked mutation   A mutation carried on one of the sex chromosomes. Sex-linked traits on the X chromosome are inherited from carrier mothers, and sex-linked traits on the Y chromosome from fathers.

Sickle cell anemia (or sickle cell disease)   A serious blood disorder caused by a single-letter genetic misspelling in the gene that produces the protein for hemoglobin. This recessive disorder leads to episodes of pain and damage to tissues and vital organs, including the retinal cells of the eyes, and, in severe cases, to premature death.

SNPs   Single nucleotide polymorphisms (or "snips"), a variation in DNA sequence occurring whenever one or more nucleotides—A, T, C, or G—in the genome differs from one individual to another.

Somatic cell gene therapy (or gene transfer)   A technique for altering genes present in the somatic (or body) cells of an individual. Unless the changes accidentally affect sex cells, they are not capable of being passed on to subsequent generations.

SSRIs   Selective serotonin reuptake inhibitors (like Prozac) that alleviate depression by slowing the removal of serotonin from regions between the neurons, leaving more of the neurotransmitter in place.

Stem cells	Relatively undifferentiated cells that retain the ability to renew themselves indefinitely through cell division and to differentiate into a wide range of specialized cell types.
Tay-Sachs disease	A fatal neurodegenerative disorder that leads to death in early childhood. It is more prevalent in eastern European Jewish populations than in others.
Transgenic animal	An animal that has had foreign DNA stably integrated into its genome.
Transhuman	"Beyond the human." Transhumanists believe that humanity is called on to improve its biological and cognitive capabilities through the use of genetic and other technologies.
Vectors	The delivery vehicles (usually viruses) used to carry corrected gene sequences into cells in gene therapy.
Veil of ignorance	A concept offered by the Harvard philosopher John Rawls to indicate the kind of impartiality required by moral reasoning about social justice. *See also* Original position.
VO2 max	The maximum amount of oxygen an athlete's lungs can take in.
WADA	World Anti-Doping Agency, an independent foundation created and partly funded by the International Olympic Committee.
Whole genome scanning	A method that uses powerful gene chips to detect up to 500,000 genetic variants at a time. By comparing the genomes of affected and unaffected people, researchers can pinpoint the variants associated with disease. Also sometimes called the "whole genome association" method.
X-SCID	X-linked Severe Combined Immune Deficiency, a genetically inherited, sex-linked form of immune system deficiency.
ZFN	Zinc finger nuclease. A specially engineered protein that can be tailor-made to allow replacement DNA sequences to home in on any identifiable DNA sequence.

Acknowledgments

I thank the John Simon Guggenheim Memorial Foundation for the fellowship that made it possible to write this book. Dartmouth College generously supplemented the grant, making a year of leave possible. Research time afforded by the Cohen Chair for the Study of Ethics and Human Values, which I hold, was of great value. I hope that this book honors the memory and generosity of Julian Cohen.

This book is aimed at the interested general reader, but as the notes only partly reveal, I draw on a great deal of scholarship and thinking about the science and ethics of human genetic engineering. The following scholars and scientists have particularly influenced me. None of them can be assumed to agree with my conclusions. On the topic of human genetic engineering generally, I draw on the work of Allen Buchanan, Dan W. Brock, Norman Daniels, Daniel Wikler, Gregory Stock, Lee M. Silver, Nick Bostrom, Leroy Walters, Maxwell Mehlman, Audrey Chapman, Erik Parens, Laurie Zoloth, and Dean Hamer. On the philosophical issues raised by disease, prevention, and genetic enhancement, I have been instructed by the work of my Dartmouth colleagues Bernard Gert and Edward Berger. John Hoberman of the University of Texas helped me in thinking about gene doping in sports.

I am grateful for the time that Theodore Friedmann, Mario Capecchi, Mark Hughes, and Janet Rossant gave me as I tried to understand the rapidly developing science of biogenetics. Sue Siegel, Robert Wells, and Katie Tillman made possible my eye-opening visit to the production floor at the gene chip maker Affymetrix.

Cathy Cramer and Judy Stern are my coteachers in our multidisciplinary Dartmouth course on ethical issues in assisted reproduction. I have learned so much from them during our years of collaboration.

My literary agent, Wendy Strothman, made many very useful suggestions at the start of this project and helped see it to completion. Cyndy Brown read the manuscript with great care and assisted me in turning philosophical and scientific jargon into readable prose. I am grateful to Jean Thomson Black and Mary Pasti at Yale University Press for their enthusiasm and editorial advice.

During my year of leave, I relied on Aine Donovan, executive director of Dartmouth's Ethics Institute, to perform the myriad duties we usually share. Her work and her advice were essential to my ability to use the year so well.

My wife, Mary Jean, is the inspiration and support for everything I do. Working on her own book project, she was my companion and support throughout this year of intensive effort. She was also my partner many years ago, when, using a "recipe" we found in an article about sex selection in *Reader's Digest,* we tried to have a boy as our second child to complement our wonderful daughter. We were successful—was it science or luck?—and that experience started me thinking about the value of deliberate interventions in human reproduction.

Index